Christian Environmentalism and Human Responsibility in the 21st Century

Christian Environmentalism and Human Responsibility in the 21st Century comprises original scholarly essays and creative works exploring the implications of Christian environmentalism through literary and cultural criticism and creative reflection.

The volume draws on a flourishing recent body of Christian ecocriticism and environmental activity, incorporating both practical ethics and environmental spirituality, but with particular emphasis on the notion of human responsibility. It discusses responsibility in its dual sense, as both the recognized cause of environmental destruction and the ethical imperative of accountability to the nonhuman environment. The book crosses boundaries between traditional scholarly and creative reflection through a global range of topics: African oral tradition, Ohio artists off the grid, immigrant self-metaphors of land and sea, iconic writers from Milton to O'Connor to Atwood, and Indigenous Canadian models for listening to the nonhuman Mother of us all. In its incorporation of academic and creative pieces from scholars and creative artists across North America, this volume shows how environmental work of its nature and necessity crosses traditional academic and community boundaries. In both form and orientation, this collection speaks to the most urgent intellectual, physical, social, and spiritual needs of the present day.

This book will appeal to scholars, researchers, and upper-level students interested in the relationship between religion and environment, ethics, animal welfare, poetry, memoir, and post-secularism.

Katherine M. Quinsey is Professor Emerita in the English Department at the University of Windsor, Ontario, Canada.

Routledge Explorations in Environmental Studies

Urban Aerial Pesticide Spraying Campaigns
Government Disinformation, Industry Profits, and Public Harm
Manuel Vallée

Addicted to Growth
Societal Therapy for a Sustainable Wellbeing Future
Robert Costanza

Coastal Disaster Risk Management in Bangladesh
Vulnerability and Resilience
Edited by Mahbuba Nasreen, Khondoker Mokaddem Hossain and Mohammed Moniruzzaman Khan

A History of Radioecology
Patrick C. Kangas

Satire, Humor, and Environmental Crises
Massih Zekavat and Tabea Scheel

UN Human Rights Institutions and the Environment
Synergies, Challenges, Trajectories
Sumudu Atapattu

Myth and Environmentalism
Arts of Resilience for a Damaged Planet
Edited by Esther Sánchez-Pardo and María Porras Sánchez

Christian Environmentalism and Human Responsibility in the 21st Century
Questions of Stewardship and Accountability
Edited by Katherine M. Quinsey

For more information about this series, please visit: www.routledge.com/Routledge-Explorations-in-Environmental-Studies/book-series/REES

Christian Environmentalism and Human Responsibility in the 21st Century

Questions of Stewardship and Accountability

Edited by Katherine M. Quinsey

Routledge
Taylor & Francis Group
LONDON AND NEW YORK

First published 2024
by Routledge
4 Park Square, Milton Park, Abingdon, Oxon OX14 4RN

and by Routledge
605 Third Avenue, New York, NY 10158

Routledge is an imprint of the Taylor & Francis Group, an informa business

British Library Cataloguing-in-Publication Data
A catalogue record for this book is available from the British Library

Library of Congress Cataloging-in-Publication Data
Names: Quinsey, Katherine M.
Title: Christian environmentalism and human responsibility in the 21st century : questions of stewardship and accountability / edited by Katherine M. Quinsey.
Description: London ; New York : Routledge, 2024. |
Includes bibliographical references and index.
Subjects: LCSH: Human ecology—Religious aspects—Christianity. | Criticism. | Nature—Effect of human beings on. | Ecotheology. | Christian ethics. | LCGFT: Literary criticism.
Classification: LCC BT695.5 C47225 2024 (print) | LCC BT695.5 (ebook) | DDC 261.8/8—dc23/eng/20230819
LC record available at https://lccn.loc.gov/2023025837
LC ebook record available at https://lccn.loc.gov/2023025838

ISBN: 978-1-032-43090-4 (hbk)
ISBN: 978-1-032-43313-4 (pbk)
ISBN: 978-1-003-36674-4 (ebk)

DOI: 10.4324/9781003366744

Typeset in Times New Roman
by Apex CoVantage, LLC

For Deborah C. Bowen
scholar and friend

Contents

Contributors

Deborah C. Bowen (Emerita, Redeemer University, Ancaster, Ontario, Canada) retired in 2017 as chair of English at Redeemer University. She continues to teach part-time, including courses in Environmental Literature and Contemporary Canadian Poetry, and is currently working with a Social Sciences and Humanities Research Council grant to produce a curated anthology, *Poetry in Place*. This, together with her previous SSHRC-funded project on "The Voice of Environmental Hope in Contemporary Ontarian Poetry," constitutes a radical re-imagining of the practical role of poetry in geopolitical place, through combining the work of ecopoets and academics of the Hamilton region, from both settler and Indigenous communities, with communal engagement in various forms. Dr Bowen's monograph *Stories of the Middle Space: Reading the Ethics of Postmodern Realisms* came out in 2010 with McGill-Queen's University Press.

Liane Miedema Brown is a PhD candidate at the University of Guelph, Ontario, Canada, and an emerging artist and writer. She uses a variety of media (from acrylic, encaustic and watercolour paints, to creative writing and scientific research) to explore the interwovenness of the natural world, drawing inspiration from her observations of nature and her own scientific research on the effects of land mosaics on forests and human-environment relationships. She is constantly seeking to understand and translate the connections between living things and the human role within the natural world.

Daniel Coleman (Professor, Department of English and Cultural Studies, McMaster University, Hamilton, Ontario, Canada) has long been fascinated by the poetic power of narrative arts to generate a sense of place and community, critical social engagement and mindfulness, and especially wonder. Although he has committed considerable effort to learning in and from the natural world, he is still a bookish person who loves the learning that is essential to writing. He has written scholarly books about literature, masculinity, migration, and whiteness in Canada, and he has written literary non-fiction books about his upbringing among missionaries in Ethiopia, about the spiritual and cultural politics of reading, and about eco-human relations in Hamilton, Ontario, the post-industrial city where he lives.

Karen Dieleman is Professor of English at Redeemer University. She writes on Victorian women's poetry, primarily as it intersects with religion and theology.

Examples include her monograph *Religious Imaginaries: The Liturgical and Poetic Practices of Elizabeth Barrett Browning, Christina Rossetti, and Adelaide Procter* (Ohio University Press, 2012), as well as journal articles on Emily Pfeiffer's evolutionary thought, Christina Rossetti's sacramental memorializing, Barrett Browning's ecotheology, and the chapter here on Alice Meynell. She plans to explore Barrett Browning's theology of work next.

Ben Faber is Associate Professor of English at Redeemer University, Ancaster, Ontario, Canada. He has presented and published papers on George Herbert, John Milton, and Andrew Marvell. While serving as Director of Teaching and Learning at Redeemer, he also presented workshops on pedagogy.

Erin Goheen Glanville is Lecturer in the Coordinated Arts Program at the University of British Columbia, Canada. Dr Glanville's research is at the intersection of pedagogy, community-engagement, and narrative arts in the field of cultural refugee studies. She produces educational media under the title Worn Words (IG: @wornwords_renarrativemedia). Dr Glanville is on the Executive Committee of UBC's Centre for Migration Studies and on the Board of Directors for Kinbrace Community Society. Recent publications include "Refugee Narrative Pedagogy" in *The Routledge Handbook of Refugee Narratives* and "Labels, Discourse, Meaning-Making" in *In/to Canada: Forced Migration from Colonization to Refugee Resettlement* (MQU Press).

Norm Klassen teaches English at St. Jerome's University in the University of Waterloo, Ontario, Canada. He is co-author of *The Passionate Intellect: Incarnational Humanism and the Future of University Education* and author of two books on Chaucer, including *The Fellowship of the Beatific Vision*, which won a 2017 Association of Catholic Book Publishers Award for new work in theology. His *Rationality Is . . . the Essence of Literary Theory* has just been published with Cascade Books.

Samuel Martin is a high school English teacher, creative writing professor, and author of the story collection *This Ramshackle Tabernacle* (Breakwater 2010), shortlisted for the BMO Winterset Award, and the novel *A Blessed Snarl* (Breakwater 2012), nominated for the International Dublin IMPAC Award. In 2018, he won ScreenCraft's Cinematic Short Story Award, and his new novel *When the Dead Are Razed*, a literary thriller, was published by Slant Books in September 2021. Samuel Martin lives in Iowa, USA.

George Piggford, C.S.C., is Professor of English at Stonehill College in Easton, Massachusetts, USA. He has published work on American and British modernism, Catholic literature, and especially Flannery O'Connor and her engagement with philosophers and mystics, including St. John of the Cross, Friedrich von Hugel, Martin Buber, and Slavoj Žižek.

Katherine M. Quinsey is Professor Emerita in the English Department at the University of Windsor, Ontario, Canada. She has published widely on Alexander Pope, John Dryden, Restoration drama and poetry, and Canadian poet Margaret Avison. Her most recent book is *Animals and Humans: Sensibility and Representation 1650–1790* (Oxford University Studies in the Enlightenment, 2017). She is currently completing two books on Alexander Pope: one on Pope and the religious imagination, and one on Pope, environmentalism, and animal welfare.

Doug Sikkema is Assistant Professor of English and the Core Program at Redeemer University, Ancaster, Ontario, Canada. He researches contemporary American literature, with a particular focus on the ways in which the material world is imagined by Christian writers in a postsecular culture, including in his studies the poet Christian Wiman and novelists Marilynne Robinson and Wendell Berry. Doug is working on a book on the postsecular Christian imagination. Outside his research, Doug edits and writes creatively. Together with their four children and two retrievers, Doug and his wife Vanessa live on a small acreage outside of Hamilton, where they aspire to be small-scale farmers.

John Terpstra is a poet and woodworker. He has published ten books of poetry, most recently, *Call Me Home*, as well as several works of non-fiction, including *Falling into Place* and *Daylighting Chedoke*, both on themes of nature and the city where he lives, Hamilton, Ontario, Canada.

Tina Trigg is Associate Professor of English at The King's University in Edmonton, Alberta, Canada, who specializes in contemporary Canadian literature and the works of Margaret Atwood. Her other research interests include inclusive education, psychology and literature, and narrative as a site of hope.

John Van Rys (Associate Professor, Redeemer University, Ancaster, Ontario, Canada) lives on a hobby farm outside Dunnville, Ontario, with his wife, two youngest children, dogs, cats, horses, and free-run egg-laying hens. When he's not at home caring for animals, he spends his time as an English professor at Redeemer University in Hamilton, teaching literature and writing courses. And when he's not teaching, he's writing—including poems, stories, essays, and textbooks. He is the author of *Moonshine Promises* (Wipf and Stock, 2021), a set of linked stories in which "Bandits" appears. He is also the lead author of *The College Writer* (Cengage Learning), and his scholarly work focuses on Canadian authors such as Margaret Avison, Alice Munro, and Carol Shields.

Susan Emily VanZanten is Senior Research Professor in the Humanities at Valparaiso University, Indiana, USA, where she served as Dean of Christ College—the Honors College, from 2018 to 2021. Over the course of 40 years in higher education, she has served as a Professor of English at Seattle Pacific University and Calvin University, as well as participated in the Lilly Fellows Program, in which she currently is a senior mentor in the Faculty Fellows Program. With a PhD from Emory University (1982), Dr VanZanten is the author of many peer-reviewed and general essays, as well as nine books in areas as varied as African literature, American literature, and Christian Higher Education. Now based in Seattle, she reads extensively, hikes regularly, and departs frequently to travel the world.

Matthew Zantingh is Associate Professor of English Literature at Briercrest College & Seminary in Caronport, Saskatchewan, Canada. He studies and teaches North American literature of the past two centuries, environmental literature, film studies, and graphic novels.

Acknowledgements

This book owes its existence, first, to George Piggford, C.S.C., who suggested the idea to me at a meeting in Ottawa, Canada, a few years ago. It has been brought to birth through the generosity, grace, scholarship, and imagination of all its contributors. Special thanks are owed to the creative artists who kindly shared their work: John Terpstra, John Van Rys, Sam Martin, Rein and Margo Vanderhill, and Rick Hill. Warm thanks to Gaspereau Press for allowing the reprinting of Terpstra's poems "Ceremony," "Giants," and "Invasive Species," to Wipf and Stock Publishers for allowing the reprinting of John Van Rys's short story "Bandits," and to Vernon Press for allowing publication of another version of George Piggford's essay "Laudato Sí, 'The Mass on the World,' and Flannery O'Connor's Eucharistic Ecology." I owe great thanks as well to the University of Windsor, its excellent library resources, and to my colleagues and former students in the Department of English and Creative Writing. Finally, I cannot say enough about the kindness, intelligence, and efficiency of the editing team at Routledge Publishing, and the helpful and incisive comments of the readers of the original proposal.

This collection is inspired by the life and work of Professor Deborah C. Bowen, a gifted teacher and inspirational scholar whose work on poetry and environmentalism, crossing boundaries of art and life, embodies the transforming power of art, reflection, and action together. To her this volume is dedicated, as a witness to the urgency and effectiveness of work like Deborah's in today's world.

All the writers contributing to this volume are deeply aware of and grateful for the land they live and work on, and the First Peoples for whom the land is home, who have been her caretakers for thousands of years. At Windsor, at the mouth of the Detroit River and amid the grasslands of the Erie basin, we recognize daily the land of the Three Fires Confederacy—the Potawatomi, the Odawa, the Ojibwa—and the Huron/Wendat people of its past. Above all, we all give thanks for the unutterable gift of the land herself, who speaks continually to us of the Creator, if only we would listen.

1 Introduction

Christian Environmentalism and Human Responsibility

Katherine M. Quinsey and George Piggford, C.S.C.

Even in a broadly secular world, the global crisis of environmental degradation evokes religious responses. These are rooted in the notion of a sentient and autonomous nonhuman world, and a sense of human guilt and responsibility for that world's suffering; they come from a variety of faiths and cultures, particularly the Indigenous spiritualities and traditions that speak most directly to these issues. Christian ecocriticism, on the other hand, occupies a complex and arguably compromised position within industrialized Western societies, whose economies are historically rooted in Judaeo-Christian concepts of human "dominion" over the nonhuman,[1] and in an ontological separation of spirit from matter, human from nonhuman, specific to Western intellectual history.

Partly in response to the implied limitations of this perspective, Christian ecocriticism and environmental activity have developed rapidly with the flourishing of ecocritical scholarship and community action in the past 50 years. Christian environmentalism is hardly new. It draws on a long tradition of environmental sacramentalism, manifested in the modern era in the work of twentieth-century mystics and theologians such as Thomas Merton, Teilhard de Chardin, Thomas Berry, and Joseph Sittler, and reaffirmed in recent years by Church leaders across both global and denominational divides.[2] Indeed, it could be argued that many of the well-recognized principles of ecocriticism themselves resemble Christian theology and spirituality. Ecocritical perspectives reaffirm the vital union of radically unlike natures that is at the heart of Christian incarnational theology. The idea that Christ's salvific work applies to all creation, human and nonhuman, has a long and orthodox history.[3] Many key principles of ecocritical awareness mirror those of contemplative spiritual disciplines: the breaking down of the subject-object dynamic, the rejection of the myths of objectivity and human separateness, and immersive listening and humility as the appropriate response to the Other.[4] Finally, this process of attentive listening is key to both environmental and social justice in the deepest sense.[5] Environmentalism by its nature entails the necessary integration of contemplation and action, theory and practice: it works through a process of conversion, in which changed vision results in communal action.[6] This suggests that effective environmental action can only be brought about through a radical inner "conversion"—defined as a transformative alteration of vision that fuels commitment to action—a process in which religious faith has a significant role to play.[7]

DOI: 10.4324/9781003366744-1

Christian environmentalism also highlights the fact that Christianity itself is not coterminous with Western culture. Many of the forces oppugning ecological consciousness are rooted not only in Western notions of human (notably European) entitlement but also in a dualistic separation of body from spirit originating in early controversies and ingrafted into Western Christianity,[8] with direct application to views of the nonhuman as ontologically distinct and inferior. As Denis Edwards remarks, in the Christian East the recognition of the threefold relationship of God, humans, and wider creation has been maintained, but not in the West, which has focused almost entirely on human redemption and relationship with God (*Deep Incarnation*, 2019, 12–13). Edwards points out that Western tradition has kept incarnational and salvific theology separate from environmental concerns, whereas these concepts are in fact central to a fully Christian environmental vision:

> The problem with the Western church's focus on redemption is not its concern with salvation, but that it has too often limited itself to human salvation, often in a highly individualistic way. What is needed is . . . an enormous extension in the recent Western understanding of [salvation in Christ], so that, faithful to the biblical promises of a new heavens and a new earth, salvation can be seen to involve the whole creation.
>
> (13)

Even within mainstream Western church tradition, however, the ecological consciousness is central and consistent. As Lynn White points out in the latter half of his famous article, the notion of "dominion" as human entitlement and right of use can be turned instead to a Franciscan understanding of dominion not as a right but as a responsibility, an obligation not to exploit nonhuman nature but to care for it. Indeed, early tenets of Christian orthodoxy, formulated in response to Gnostic philosophies that attempted to separate spirit from matter and God from creation, expressed fundamentally ecological values: the essential goodness of the physical world, its diversity as evidence of God's love and a reminder of the limitations of human knowledge, God's immanent and active presence in each element, and human responsibility to care for the creation. These ecological principles can be traced from medieval concepts of the Book of Nature as expressive of God's being and purpose through to early modern and Romantic evangelical ideas of animal welfare and humanitarianism, rooted in imaginative relationship to the nonhuman world. The belief in the millennial redemption of the nonhuman creation and its inhabitants, that their primary purpose is not merely to serve humans but to glorify God in their own natures, appears consistently from early patristic writings through the Enlightenment to the present day.[9] The idea that humane treatment of the nonhuman is itself a sacramental as well as a civic duty, reflecting God's own care for the nonhuman creation, can be found both in early modern sermons and in recent theological statements by mainstream religious leaders across both Eastern and Western Christianity. Biblical and historical studies show that animal welfare and ecologically centred principles were the biological, social, and spiritual heart of the ancient agrarian society that linked these practices to individual and collective spiritual and physical health, as well as to human social welfare.[10]

Finally, in both academic and community-based settings, much contemporary Christian ecocriticism highlights the intersection of nonhuman environmental justice with human social justice. Joshua Mabie sees the turn away from the anti-humanism of early deep ecology towards a postcolonial ecocriticism tied to human social justice as a process in which Christian environmentalism has a specific role to play:

> Ecocriticism's heightened attention to ways that environmental problems threaten the lives and livelihoods of the world's poorest and most vulnerable people provides an opportunity for Christians to become involved in both the scholarly project of ecocriticism and in the work of advancing global environmental justice . . . ecocriticism's postcolonial turn has also foregrounded the needs of hungry, thirsty, exiled, naked, sick, and imprisoned people around the world, those Fanon described as 'the wretched of the earth' and those that Christ calls 'the least of these.'
>
> (Mabie, "The Field is Ripe," 2016, abstract; 288)

In an afterword to the same issue of *Christianity & Literature*, renowned ecocritic Lawrence Buell supports Mabie's claims and links them to those of Pope Francis's seminal 2015 encyclical:

> By no coincidence do the terms of his argument resonate closely with the 'integral ecology' that informs Pope Francis's 2015 encyclical *Laudato Sí*, itself so strongly influenced by Latin American liberation theology. 'A true ecological approach,' Francis sums up, must 'hear both the cry of the earth and the cry of the poor.' He then proceeds, as it were, to call in the 'ecological debt' owed by the rich nations of the global north to the global south and demands that it be paid.
>
> (Buell, "Afterword," 2016, 365, quoting *Laudato Sí. On Care for Our Common Home*, 2015, 22)

Another devastating "ecological debt" is the one incurred in the interaction between European settlers and the Indigenous inhabitants and environment of the New World, in which early modern Europeans (self-designated as "Christians") claimed the right of divine entitlement to utilize the land, in the manner described by Lynn White, and assigned themselves superiority over the existing human inhabitants. The association of the Church with these depredations both ecological and social (in the Canadian residential schools designed to assimilate Indigenous children into settler society, for example) continued through to the late twentieth century. As suggested in some of the contributions to this collection, in particular Daniel Coleman's reflection and John Terpstra's poetry, humility and spiritual openness to the land as its own place, and a restoration of ancient contemplative disciplines, parallel to Indigenous ways of knowing the nonhuman creation, are key to both environmental healing and human social healing and justice.

The specific focus of this collection is on this notion of stewardship or ecological accountability in all its senses. The questions around the term "stewardship" itself are well framed in a recent review by Anthony Milligan: its fundamental hierarchy and anthropocentrism, and the failure of humans to live up to the role not least.[11] "Stewardship" as a term invokes multiple ironies. For example, it is a term used to describe both financial giving in Christian churches and the fundamental relationship of Indigenous peoples to the land (often acknowledged as their "stewardship" by White settler institutions). In the specifically Christian theological application of the term, however, stewardship as "Creation Care" means *human responsibility* for the wellbeing of the nonhuman creation. Both terms are used in this book's title, as equally double-edged. "Human responsibility" can refer both to the imperative to humans to care for the nonhuman creation (and each other in so doing) and, at the same time, to the increasingly evident anthropogenic causes of recent environmental degradation. The term also raises questions around anthropocentrism, both that of traditional Abrahamic theology, which sees humans as the apex of Creation, and that of human power to destroy the environment, which has no partner in the nonhuman animal world—a form of exceptionalism we could well do without. In the face of climate change and species loss, attempts to restore the balance of nature are now of necessity anthropogenic, involving human interposition. As Milligan points out (review of Melin, 2018, 255), the human claim to stewardship can be seen as fundamentally absurd (we were "absentees" for most of the history of the earth) but contemporary facts point to its necessity.

In the current climate of opinion, then, not to mention the physical climate of the planet, human responsibility is the overwhelming fact and focus, both scientifically supported and popularly contested. Christian ecocritical scholarship speaks directly to that question, and to many of those raising it. It centres on the dual sense of human *responsibility*, both as being the direct cause of catastrophic global change and as being accountable to the nonhuman environment, having an ethical and results-oriented obligation of care—an obligation that is key not only to our own physical survival but also to our core human identity and spiritual health.[12] As suggested by Pope Francis in *Laudato Sí*, the will to accountability and the power to act on it come not from a position of separateness and superiority but from one of deep humility and engagement with, and within, the nonhuman environment and its other-than-human denizens. This is the same logic applied in the Gospel to human social justice—in which those who are the poorest and most marginalized are specifically those with whom our Lord identifies (Matthew 25:40–45), a model that shares with radical environmentalism a deep challenge to human divisions of hierarchy and power, and even to categories of being. In both theoretical and practical terms, then, an environmental activist perspective is entirely consistent with Christianity: in the assumption of human limitations; in the value of humility and receptiveness, central both to a relationship with God and to an ecocritical awareness of the nonhuman world; and in the urgency of the call to social justice, which for Christians is both immediate and eschatological.

This volume focuses specifically on this view of human responsibility. It draws from an extensive body of Christian ecocritical scholarship, not only the specific strong thread within environmental scholarship of the past 60 years but also the roots of that perspective in Christianity itself, in which a fundamentally ecological understanding is inherent to the core of Christian incarnational theology, extending even to the spiritual and moral self—the humility and listening that are key to our relationship with God are also key to our relationship with the nonhuman creation through which God speaks. To be a Christian is to offer witness to the divinity inherent in the natural world and all its creatures, seeing in their autonomy and variety—so under threat in today's world of vanishing species—the multiplicity in which God delights. It is to understand, in the words of the Ecumenical Patriarch Bartholomew, that "the divine and the human meet in the slightest detail in the seamless garment of God's creation, in the last speck of dust of our planet" (Closing Remarks, Halki Summit I, Istanbul, 20 June 2012, qtd in Francis, *Laudato Sí,* no. 9). In *Laudato Sí* Pope Francis has reminded us that "creation" in the Christian context has a broader meaning than "nature," "for it has to do with God's loving plan in which every creature has its own value and significance" (no. 76). Leaders in The Evangelical Climate Initiative have noted that climate change disproportionately harms those least able to cope with its effects:

> Poor nations and poor individuals have fewer resources available to cope with major challenges and threats. The consequences of global warming will therefore hit the poor the hardest, in part because those areas likely to be significantly affected first are in the poorest regions of the world.
>
> (Climate Change: An Evangelical Call to Action, 2006)

Environmentalism provides Christians with an inspirational and immediate—and increasingly urgent and practical—set of frameworks within which to fulfil the basic commands of their faith: to love God and one another. Conversely, those basic commands themselves point the way to ecological wholeness for the global community, both human and nonhuman.

Christian Environmentalism and Human Responsibility in the 21st Century comprises original scholarly essays and creative works, exploring the implications of Christian environmentalism through cultural criticism and creative reflection. Our collection ranges through academic essays, poetry, memoir, and short story, from an array of leading Canadian and American scholars and creators, a generic variety intended to speak to the immersive nature of this issue and the immediacy with which we need to respond—a response that transcends traditional boundaries of theory and practice, academy and broader community, faith, faiths, and secularism.

The collection contains some critical studies from a global context, and some on both American and British subjects, both contemporary and historical, as well as reflections on Canadian literature and popular culture. It features poetry, memoir, and reflection rooted in specific places and their layers of being. Environmental writing is deeply about place; the local speaks to the universal. The collection thus contains a cluster of pieces creative, critical, and reflective, about the unique geological and social ecosystem of Hamilton, Ontario, an industrialized urban community cohabiting with an ancient Indigenous community and a breathtaking geological formation that speaks its own narrative. The local is universal: here, in a specific place, are focused key environmental concerns—living ecologically; layers of geological history; nonhuman inhabitants; human history and communities both settler and Indigenous, and their relationship to the nonhuman environment; nonhuman inhabitants; responses of individuals and communities in action and reflection.

More deeply, this collection emphasizes the environmental response in literature and culture, in both creative works and critical studies—the recording of human experience of the nonhuman, the "attentiveness" to the detail of place, and the immersive engagement with the nonhuman other as an act of immersion with the ultimate Other whose Self creates us as our own truest selves. This is also the act of listening to the nonhuman other that speaks to us, a recognition of the language that pours forth outside human interpretation and imposition. As suggested in Sam Martin's memoir, art is redemptive and transformative. This is the dynamic that can bring about the fundamental change in human vision, positionality, and humility towards the nonhuman Other and to the One who gave it being, that empowers human action and responsibility.

The collection opens with a reflective essay by Liane Miedema Brown, who "explores our eroding identity as members of an ecosystem, and our lost sense of place." This creative meditation addresses the core principle that while Christians look to a home beyond this world, they have a responsibility to recognize their own being as creatures made from the soil, members of a sustaining ecosystem, and as participants in a community both social and biological. In order to bring about a righteous world, one must live into the realization that "our earthiness is not a taint but [rather] the foundation from which we build the Kingdom of God." She encourages readers to "realize [their] place," both metaphorical and physical, by the practice of "lavishing attention" on the created world around them and on God's permeating presence within it.

The poetry of noted Hamilton-area ecopoet John Terpstra helps shape this collection, as both comment and reflection, at intervals between the prose essays and stories.

Two ecocritical essays draw on cultural and global perspectives. One, by Susan Emily VanZanten, performs an ecocritical reading of the African oral work *Jesus of the Deep Forest*, produced in 1971 by Afua Kuma, a non-literate Ghanaian village woman. Composed in the Ghanaian genre of *apae*, or high poetry of praise, *Jesus of the Deep Forest* was originally recorded and transcribed in the Akan language in the late 1970s and was translated and published in print in 1981. As "an extended song of praise in honour of Jesus as the ultimate political and spiritual ruler within the context of a gradually modernizing African agricultural society," the poem expresses the ecological harmony of an ancient culture living within the context of tension between sustainability and agricultural/technological development. This poem thus challenges the Western perspectives, which locate destructive models of human dominion within the Judaeo-Christian framework (such as that expounded by Lynn White) in its "depiction of permeable boundaries between divine transcendence and the natural world and in its account of the divinity inherent in the natural world and its creatures because of God's creative work." Jesus is celebrated both as a traditional chief and as the power bringing life to people, animals, and the land alike.

Erin Goheen Glanville addresses the current discourse on migration and displacement, in which forced migration, diaspora, refugee lives, and human movement more generally, as well as climate crisis, are often imaged through water and seascape, both literally and figuratively, to the extent that water used in this way has become a dead metaphor, both loosed from and entrapping its referent in attributed significations. Through reading the graphic memoir *The Best We Could Do* by Thi Bui, in which Bui reflects on and reimagines the refugee experience of her own Vietnamese family, Glanville shows how Bui's images reinvigorate the water metaphor's dynamic of difference and open up its reference, while layering the human and nonhuman bodies so that the nonhuman speaks through both. Bui thus creates a fresh metaphor for refugee lifeworlds while freeing both human and nonhuman from limiting interpretations.

Two essays providing ecocritical perspectives from early modern literature and culture show the historical and cultural movements through which Christian ecocritical concepts emerge, and the extent to which they are applicable to the contemporary work of both the classroom and of broader social movements such as animal rights. Ben Faber explores how an ecocritical reading of Milton's *Paradise Lost* can operate effectively in a contemporary undergraduate classroom, by approaching directly a key question raised by the poem, that of the apparent punishment of the nonhuman world for humanity's sin in the Fall. Through "meeting Milton's expectations for the implied reader of *Paradise Lost*," students can engage themselves in the moral and ethical challenges presented both in the poem and in the current environment of anthropogenic climate change. An ecocritical approach thus "allows the epic itself to become a way of probing the dominant [Western] narrative, of developing practical critical skills, and of encouraging divergent thinking about paradises lost and reclaimed."

Katherine M. Quinsey focuses on post-Reformation concepts of the theology of animal welfare, with particular attention to the sermon literature of the late seventeenth century. The influence of classical scepticism and post-Reformation theology, together with the impact of the new science, combined to produce a strong thread within seventeenth-century theological writing that saw nonhuman animals as autonomous beings separate from human definition, created not for the service of man but to glorify God in their own natures, as "fellow creatures" with humans. This is a notable thread in Anglican sermons and religious writing of the seventeenth and eighteenth centuries. Focusing on seventeenth-century clergyman Thomas Hodges, the essay argues that the language and theology of Hodges's sermons vividly express a new sense of animal identity, sentience, and status before God, drawing both on the scepticism of Plutarch and on Biblical exegesis, in a manner that anticipates not only the language of late eighteenth-century humanitarianism and its strong links with Calvinist tradition but also the notion of animals as having legal rights, almost 150 years before animal welfare legislation was first enacted in Britain.

The short stories of John Van Rys take place in the quintessential nexus of urban and rural values, human and nonhuman relationships, in the setting of the hobby farm, where former urbanites care for nonhuman animals in the light of contemporary environmental consciousness. The story reprinted here, "Bandits," about the attempts to humanely prevent raccoons from destroying a chicken (and egg) run, delightfully and wittily explores the dynamics of human control and definition with creaturely autonomy, and the complexities of creaturely care and human responsibility, within the context of a larger redemptive and common perspective. Van Rys's story is accompanied by a reflective essay showing how his "creative work struggles with these issues 'on the ground' . . . in concrete narratives and images rather than in theory; [his] discussion [seeks] to understand biblical inspiration for creature care as part of creation care."

Karen Dieleman writes about the nineteenth-century Catholic poet Alice Meynell in a landmark essay and evocative reading linking Meynell's deeply incarnational theology with her ecological perspectives. Using Tim Ingold's concept of an immersive engagement, of living as a wayfaring among "things" with their own being, rather than travelling past objects, Dieleman shows how this concept is central not only to the poet's own vision but also to her concept of Christ. "Meynell considers the incarnated, crucified, and yet living Christ as continuing to inhabit or wait within or walk among the grain, grass, grapes and galaxies of the world, devoted to their particularity." As Christ marks the creation through His coming and going, being and becoming within it, the creation is seen to be a meshwork, a topography both spiritual and material, which by gift of language and vision continually surprises within the ordinary.

George Piggford explores the relation between Teilhard de Chardin and Flannery O'Connor in their environmental and Eucharistic visions, and the ways in which they anticipate the approach advocated by Pope Francis in *Laudato Sí*, his landmark 2015 encyclical on the interconnectedness of all living things, human and nonhuman. The essay's main literary focus here is "A View of the Woods," published in 1957, in which O'Connor not only emphasizes the autonomy and the indwelling divinity of the natural world but also envisions the current environmental degradation and the anthropocentric greed that underlies it; indeed, she envisions our own planet, and all life upon it, as the wounded and bleeding Christ, especially in the face of industrialization and exploitation aided by modern technology. Accordingly, "there is clear evidence that for both the current Pope and Flannery O'Connor our global environmental crisis reflects and results from a deeper spiritual crisis and requires an 'integral ecology,' one that is concerned with cultural structures, economic justice, environmental awareness, and spiritual development" (Francis 2015, 104 [no. 137]), a vision in which social and environmental justice are integrally related, calling for a "conversion" to full awareness of our connection to all creatures.

In "'Can you make this all run again?' The Art and Environmentalism of Margo and Rein Vanderhill," Samuel Martin pens a deeply moving memoir and reflective essay around the story of two artists in his small Iowa community, whose work and practical lives reveal

> environmentally conscious, creative ways to love a place and its people, even when that place seemed to turn on them in the fall of 2018 in a flood that nearly destroyed their house and over fifty years of collected artwork.

His vision reflects the influence of Don MacKay's notions of the self's relation with place, of "home" and place and its otherness, not to be appropriated by the self but rather of finding the self in the act of listening through "poetic attention" that recognizes, respects, and values the other's wilderness. In the work and life of the Vanderhills, this attention to place bears witness to the actualities of their home, and to the ongoing unsettled relation between cultivation and nonhuman nature. In the tiniest of details, there is witness to a life that is separate, both familiar and unfamiliar, a homely otherness beyond categories of human definition and perception. "Such attention, according to McKay, can make possible 'a *translation*' as opposed to 'a *vestige*' of this place—a flawed, human attempt to see more truly a part of the whole, since the whole cannot be grasped."

Norm Klassen explores the relationship between sacramental theology and environmentalism, seeing a sacramental view of reality both as preparing the ground for environmental responsibility and as "cultivat[ing] a wariness" of human pretensions, including those of secular environmentalism. He goes on to read closely the expressive work of two very different Canadian artists—singer-songwriter Bruce Cockburn and novelist David Adams Richards—in the light of this understanding, seeing the dynamic relationship of nonhuman nature and ineffable grace in Cockburn's early work, and the stripping away of hypocrisy, with the implicit demands of human responsibility, in Richards's novel *For Those Who Hunt the Wounded Down* (1993). In this book, Richards upholds a vision of creatureliness both positively in the discoveries of his main characters and negatively in his presentation of university figures, as they utilize environmentalism as a tool for other ends.

The final essays of the collection follow a selection of poems by John Terpstra to focus on contemporary nature writing, Canadian and Indigenous perspectives, and the autonomous, inspirited environment. Tina Trigg moves beyond the dystopian environmentalism of Margaret Atwood's MaddAddam trilogy to consider how Atwood's work across genres embodies a more nuanced and yet challenging environmental vision of "relationality, interdependence, and adaptation," de-centring human perspectives while laying down paths for radical human action. She moves from the early short story "The Resplendent Quetzal" (*Dancing Girls and Other Stories*, 1977), in which birding is both "catalyst and camouflage" for the relationality and adaptation that are a response to loss, to the recent graphic novel series *Angel Catbird* (2016–2017), which embodies disruption not only of species divisions in its characters but also of form, content, voice, and generational perspective. Mingling fantasy fiction with real-world instructions, aiming at a late childhood readership, it centres marginalized, less-visible relations and shows their urgency, while breaking down the barrier between literary art and real-world practice. Trigg connects Atwood's poetic attentiveness—learning to see, and the sacramental understanding of the world that entails—to her theme of birds and birding, as both metaphor and literal action or state. Through this dynamic in Atwood's writing, the human perspective is challenged and shifted literally to the bird's eye view, "following the birds" to see what is happening on the globe (Leber, *Audubon*, 2019).

Matthew Zantingh reflects on John Terpstra's trilogy of Hamilton poems—*Falling Into Place* (2002), *The House with the Parapet Wall* (2014), and *Daylighting*

Chedoke (2018)—with particular attention to the way these poems think and express the sensation of place, and what they have to offer readers in the way of revising their own notion of place. Terpstra digs into Hamilton's specific geography and its identity as "an urban place that has been deeply scarred by decades of industrialization, significant pollution, and civic neglect." His poems trace the imagery and sensation of human interactions with that geography; "daylighting Chedoke," for example, refers to visionary openings of a long-buried creek lost under layers of industrialized structures. In thinking of this place as a gift, Terpstra begins to move towards what the Haudenosaunee scholar Vanessa Watts calls "Place-Thought"—a traditional way of being within the nonhuman environment, one that presents the possibility of some redemption, and one by which Terpstra "seeks to honour all of his neighbours, whether human or more-than-human" in the place he lives.

Daniel Coleman addresses the confrontational relationship with the nonhuman world that is the basis of most European Settler culture and economy, by investigating "two much under-valued traditions" that could open up a more reciprocal relationship with the nonhuman natural world. The first of these is the Two Row-Covenant Chain Wampum agreement made in the seventeenth century between the Haudenosaunee and the European settlers on the Hudson River, "considered the foundational treaty" between the incoming Europeans and the Indigenous peoples of the northeast. The second is the *via negativa*, or apophatic tradition, of Christianity, also relatively obscured today through the long dominance of more imperialist versions of the faith. Both these traditions emphasize the need for humbly setting aside one's own knowing and self-definition in order to be known by the Other, a position of immersive listening that is also key to environmental awareness.

> Maybe, if twenty-first century people living in a period of growing environmental urgency can retune our ears, we can learn to hear what the land, what Creation, and the people who built their cultural ways through dialogue with it, might be saying.

The collection concludes with a memoir by Douglas Sikkema. Drawing from his own experience in a small farming community in the Niagara region, Sikkema "shares what it looks like to undertake healing work in the material world in the context of a lived mythology that counters many of the dominant narratives (or myths) of the secular Anthropocene." He shows that recovering a sense of the world as "a place enchanted" by God, full of God's attention and presence, embodying and revealing of God's glory, constitutes an immediate and practical call to act responsibly towards that nonhuman world.

In an Afterword, Deborah C. Bowen reflects on the interconnectedness and various voices of these essays and creative reflections, and on the urgency of action derived therefrom. The attentiveness to place and immersive listening are also reflected in her own experience, in particular her recent work, which brings together poets and community members in shared engagement.

As is evident in its dedication, this collection honours the contributions of Dr Bowen to the fields of critical theory, Canadian literature, and Christian environmentalism.

Her recent work, recipient of two Social Sciences and Humanities Research Council Insight Grants, constitutes a radical re-imagining of the practical role of poetry in place, in the created world both human and nonhuman, through a multifarious project that combines the work of ecopoets and academics of the Hamilton region, both settler and Indigenous communities, with communal engagement in various forms. In its incorporation of academic and creative pieces, from emerging to distinguished scholars and creative artists, this volume reflects the polysemous nature of such environmental work, which, of its nature and our necessity, crosses traditional academic and community boundaries, integrating the critical with the creative, the reflective with the active. Christian environmentalism itself relates uniquely to a nexus of contemporary debates around questions of human responsibility. Thus, in both form and orientation, this collection speaks to the most urgent needs—ecological and social, intellectual and spiritual—of the present day.

Notes

1 This position was famously put forward by Lynn White, Jr., in "The Historical Roots of Our Ecologic Crisis," 1967. For a recent examination of the influence of White's essay, see LeVasseur and Peterson, *Religion and Ecological Crisis*, 2017.

2 Signature works include Teilhard de Chardin, *The Divine Milieu*, 1957; Joseph Sittler, "A Theology for Earth," 1954 and *Essays on Nature and Grace*, 1972; Thomas Berry, *The Sacred Universe: Earth, Spirituality, and Religion in the Twenty-First Century*, 2020. For examples of recent commentary, see Pihkala, *Early Ecotheology*, 2017; Grumett, *Teilhard de Chardin*, 2005; Kehoe, "Thomas Merton's Ecopoetry," 2009. While de Chardin is a deeply problematic figure in many respects, his environmental philosophy is widely influential. See George Piggford's essay in this collection.

3 Seen, for example, in early commentaries on the restored Edenic harmony of nonhuman creatures in Isaiah and the redemption of "the creature" in Romans 8. Some recent writers go further, to challenge incarnational theology's traditionally anthropocentric focus: "The incarnation of God in Christ can be understood as a radical or 'deep' incarnation, that is, an incarnation into the very tissue of biological existence, and system of nature" (Gregerson, "The Cross of Christ," 205, qtd in Edwards, *Deep Incarnation*, 11–14). Mark Eaton, citing both de Chardin and Merleau-Ponty, argues that the human form should not be seen as the sole expression of God's Incarnation in the created world (Eaton, "Beyond Human Exceptionalism," 9–10, 202–217). David L. Clough argues that the notion of "the centrality of humanity to God's purposes is biblically and theologically both unnecessary and undesirable, and . . . that Christianity should be distinguished by its theocentrism, rather than anthropocentrism" (*On Animals*, 2012).

4 See, for example, Kehoe, "Thomas Merton's Ecopoetry."

5 Nicely expressed by Mark F. Cladis, "Stone-Throwers," 2008, 81–108.

6 Outlined by Pope Francis in the 2015 encyclical *Laudato Sí*, cited here by Mark Eaton, "The Environmental Imagination," 277.

7 Another locus of pressure on modern Christian environmentalism can be found in the tension between "God and Green" (Wilkinson, 2012), evident in some theological conservatism that sees today's environmentalism as a pagan religion in which nature replaces God and humanity is de-centred (Blanchard and O'Brien, 2014, 17–19), or which preaches a deterministic rejection of human responsibility as an arrogant incursion on God's care for creation (Brown, 2021). These tensions have given rise to studies by a wide range of scholars including Lin, Wilkinson, and Gottlieb, all of whom respond in their studies by incorporating the environmental responsibility imperative into varied Christian frameworks. A response can also be found in the rich tradition of Christian

environmental writing both historical and contemporary, as well as the ecologically informed theological positions adopted by mainstream Church leaders across global and denominational boundaries.

8 For brief accounts of this process, see Morton Kelsey, *Encounter with God*, ch. 4, and *Healing and Christianity*, ch. 9. See also Muddiman, "New Testament," 1998, 25.

9 See, for example, McLaughlin, "Evidencing the Eschaton."

10 For a recent study, see Ellen F. Davis's excellent *Scripture, Culture, and Agriculture*.

11 Anthony Milligan, review of Melin, *Living with Other Beings*, 2013, 225: "Christian perspectives on the non-human have been in something of a state of flux in recent decades with the notion of stewardship coming under sustained pressure for several reasons. First, it seems rather anthropocentric and hierarchical. Second, it may be too flattering for humans and therefore in conflict with the ecologically pivotal notion of humility. Finally (and relatedly), we stewards seem to have been absentees for most of the history of life on Earth, making such a role seem redundant or at least unclear."

12 Follows Buell's classical formulation in which an ecological approach is defined as one that challenges human-centred interpretations of the cosmos; that recognizes the nonhuman creation as having its own significance separate from human definition, and its own narrative separate from human symbolism; that entails both humility and responsibility towards the nonhuman environment (*The Environmental Imagination,* cited in Bryson, *Ecopoetry*, 5–6, and in Scigaj, *Sustainable Poetry*, 10). These principles are given a spiritual dimension most notably in Francis's *Laudato Sí*, especially chapter 6, parts 3 and 4.

References

Berry, Thomas. *The Sacred Universe: Earth, Spirituality, and Religion in the Twenty-First Century*. New York: Columbia University Press, 2020.

Blanchard, Kathryn D., and Kevin J. O'Brien. *An Introduction to Christian Environmentalism: Ecology, Virtue, and Ethics*. Waco, TX: Baylor University Press, 2014.

Brown, Caleb. "Acts of God and Men: Addressing the Tension Between Divine Sovereignty and Human Responsibility in Christian Ecological Ethics." *Journal of the Oxford Graduate Theological Society* 2, no 2 (November 2021): 12–22. https://jogts.org/index.php/jogts/article/view/33.

Bryson, J. Scott. *Ecopoetry: An Introduction*. Salt Lake City: University of Utah Press, 2002.

Buell, Lawrence. "Afterword." *Christianity and Literature* 65, no. 3 (2016): 364–367. https://doi.org/10.1177/0148333116631225.

———. *The Environmental Imagination: Thoreau, Nature Writing, and the Formation of American Culture*. Cambridge: Harvard University Press, 1995.

Cladis, Mark. "Stone-Throwers with Excellent Aim: Waking-Up to an Environmental Democratic Vision." *Religion & Literature* 40, no. 1 (Spring 2008): 81–108.

Clough, David L. *On Animals: Volume One—Systematic Theology*. London: T&T Clark, 2012. Kindle edition.

Davis, Ellen F. *Scripture, Culture, and Agriculture: An Agrarian Reading of the Bible*. Cambridge: Cambridge University Press, 2009. Kindle edition.

Eaton, Mark. "Beyond Human Exceptionalism: Christology in the Anthropocene." In *Religion in the Anthropocene*, edited by Celia Deane-Drummond, Sigurd Bergmann and Markus Vogt, 202–217. Eugene, OR: Cascade, 2017. Kindle edition.

———. "The Environmental Imagination." *Christianity and Literature* 65, no. 3 (2016): 276–278. Introduction to special issue. https://doi.org/10.1177/0148333116646046.

Ecumenical Patriarch Bartholomew. *Closing Remarks, Halki Summit I*, Istanbul, 20 June 2012.

Edwards, Denis. *Deep Incarnation: God's Redemptive Suffering with Creatures*. Maryknoll, NY: Orbis Books, 2019. Kindle edition.

Evangelical Climate Initiative, The. *Climate Change: An Evangelical Call to Action*, 2006. www.influencewatch.org/app/uploads/2020/08/climate-change-an-evangelical-call-to-action.-08.20.pdf. Accessed 6 May 2023.

Francis. *Laudato Sí: On the Care for Our Common Home*. Frederick, MD: The Word Among Us Press, 2015.

Gottlieb, Roger S. "Introduction: Religion and Ecology—What Is the Connection and Why Does It Matter?" In *Oxford Handbook of Religion and Ecology*, edited by Roger S. Gottlieb. Oxford: Oxford University Press, 2006, 2009.

Gregerson, Neils. "The Cross of Christ in an Evolutionary World." *Dialog: A Journal of Theology* 40, no. 3 (Fall 2001): 197–207. Cited in Edwards, *Deep Incarnation*, 2019, 13.

Grumett, David. *Teilhard de Chardin: Theology, Humanity, and Cosmos*. Leuven: Peters, 2005.

Kehoe, Deborah. "Thomas Merton's Ecopoetry." *The Merton Annual* 22 (2009): 170–188.

Kelsey, Morton. *Encounter with God: A Theology of Christian Experience*. Minneapolis: Bethany Fellowship, 1972.

———. *Healing and Christianity in Ancient Thought and Modern Times*. London: SCM Press, 1973.

Leber, Jessica. "Margaret Atwood Insists Birds Matter to Everyone—Whether They Realize It or Not." *Audubon*, 25 November 2019. www.audubon.org/news/margaret-atwood-insists-birds-matter-everyone-whether-they-realize-it-or-not.

LeVasseur, Todd, and Anna Peterson, eds. *Religion and Ecological Crisis: The "Lynn White Thesis" at Fifty*. London: Routledge, 2017. Kindle edition.

Lin, Johnny Wei-Bing. *The Nature of Environmental Stewardship: Understanding Creation Care Solutions to Environmental Problems*. Eugene, OR: Wipf and Stock, 2016.

Mabie, Joshua. "The Field Is Ripe: Christian Literary Scholarship, Postcolonial Ecocriticism, and Environmentalism." *Christianity and Literature* 65, no. 3 (2016): 279–297. https://doi.org/10.1177/0148333115585488.

McLaughlin, Ryan Patrick. "Evidencing the Eschaton: Progressive-Transformative Animal Welfare in the Church Fathers." *Modern Theology* 27, no. 1 (January 2011): 121–146. https://doi.org/10.1111/j.1468-0025.2010.01658.x.

Milligan, Anthony. Review of *Living with Other Beings: A Virtue-Oriented Approach to the Ethics of Species Protection* (Studies in Religion and the Environment; Berlin: LIT Verlag, 2013), by Anders Melin. *Journal for the Study of Nature, Religion, and Culture* 12, no. 2 (2018): 225–227. https://doi.org/10.1558/jsrnc.31577.

Muddiman, John. "A New Testament Doctrine of Creation?" *In Animals on the Agenda: Questions About Animals for Theology and Ethics*, edited by Andrew Linzey and Dorothy Yamamoto, 25–32. Chicago: University of Illinois Press, 1998.

Pihkala, Panu. *Early Ecotheology and Joseph Sittler*. Zürich: Lit Verlag, 2017.

Scigaj, Leonard M. *Sustainable Poetry*. Lexington: University of Kentucky Press, 1999.

Sittler, Joseph. *Essays on Nature and Grace*. Philadelphia: Fortress Press, 1972.

———. "A Theology for the Earth." *The Christian Scholar* 37, no. 3 (1954): 367–374.

Teilhard de Chardin, Pierre. *The Divine Milieu*. 1957. Translated by Bernard Wall. New York: Harper & Brothers, 1960.

White, Lynn Jr. "The Historical Roots of Our Ecologic Crisis." *Science* 155 (10 March 1967): 1203–1207. https://doi.org/10.1126/science.155.3767.1203.

Wilkinson, Katherine K. *Between God and Green: How Evangelicals Are Cultivating a Middle Ground on Climate Change*. New York: Oxford University Press, 2012.

2 The Practice of Lavishing Attention

Liane Miedema Brown

There are wind maps on the water's surface. I am driving 90 km/hr along the Burlington Skyway. 3 p.m. on a Thursday, and the sky is clear and blue. This bridge is one of the few places along my drive that I have a clear view of the water and, tempted by the fiery smokestacks that dominate the skyline in dramatic contrast to the calm of the water, I glance over the rail towards the Hamilton Harbour. There they are—shimmering ripples traced in curving patterns, snaking across the expanse to my right. The rest of the harbour is deep blue and calm. The only way to make sense of these trailing textures, so distinctive against the rest of the water, is if the wind travelling across the harbour is disturbing the surface just enough to reflect the wind's movement along these very specific trails. It is as if there are silver rivers travelling on top of the water, mapping out the pathway of the wind. Wind, which from my vantage point, I can neither see nor feel. But I can still see traces of the movement and interactions beyond myself. The view only lasts for a second. Self-preservation draws my eyes back to the road. I am moving too fast and there is too much traffic for me to look again before I am past the harbour. The image sticks with me. It is imprinted in my mind so much so, that when I arrive home I pull out my journal and write out those words: "There are wind maps on the water's surface."

There are so many times in my life like that moment in May, moments when I wish I could pause the world and pour my undivided attention into my surroundings, opening myself up to the wonder of the place I am living in. I want to take the time to marvel at the way water responds to its surroundings with such acceptance and fluidity. How wind moves, unseen, reflected only in water, branches, loose leaves, and discarded wrappers. There is so much life beyond myself, so many interactions peripheral to my awareness, that there is no way I can observe or understand it all. Even so, I desperately want to. I want to allow my attention to magnify and transform the mundane. How strange it is that *mundane* means both dull and uninteresting and "of this earthly world rather than a heavenly or spiritual one" (*American Heritage Dictionary of the English Language*, 2016). The earth is anything but dull. Give it the barest sliver of attention, and this earthy realm transforms. The simple things become marvellous, calling out to be noticed. My curiosity draws me in to the minute details or the grand landscape visions. The world is

DOI: 10.4324/9781003366744-2

harkening me closer. As I attend to these whispers of presence, it begins to reveal itself, quietly, layer by layer. As Daniel Coleman writes:

> Anyone who begins to pay attention, real attention, to even one square metre of any place on earth, from the microscopic beings that turn leaf matter into soil to where water goes after soaking into that ground, will soon be overwhelmed.
> (Coleman, *Yardwork*, 2017, 10)

And it is true! I am overwhelmed daily, constantly delighted by simple things. I marvel at the capacity of the Yarrow and Queen Anne's Lace to root themselves into the smallest cracks in the sidewalk. Their tenacity and resilience are what earns them the title *weeds*, but their bursts of tiny frothy white flowers against the grey pavement and green of boulevard grasses are as exquisite as any Lilac or Baby's Breath. I have learned where the local mulberry trees are, how to look out for the stained purple footprints on the sidewalk, and the tinted bird droppings on the cars to know that the berries are beginning to ripen. I watch the swelling paper covers at the end of spruce branches to harvest for the sweet taste of spruce soda in the springtime. Thanks to an equally enthusiastic friend, I know where the decorative golden raintree grows and when the pods blush pink and orange in the summer.

Simone Weil wrote that "attention is the rarest and purest form of generosity" (Pétrement, *Simone Weil*, 1976, 486), and I am inclined to agree. This feels generous. Lavishing. Like a muscle, the more I flex my attention the more easily it comes. Now, it is nearly impossible for me to avoid moments of wonder, pausing mid-stride to pat the minuscule fronds of moss growing in mounded cushions along a stone wall, or stopping to take a picture of the single droplet of water clinging to a service berry tree, opalescent against the grey sky on a rainy day. The sky and the clouds and the colours of the new leaves on the trees, the way that grass moves with the wind, the sound of birds in spring—I am, constantly and continually, falling in love with the world. Lavishing attention on every pocket of green, grey, or brown in this city. I pour my mind and heart over the world like an anointing oil. An acknowledgement of something grand and holy. Attention does not create connection and oneness with the world around me. In the same way that oil is a signal of something already present—a physical embodied acknowledgement of the holy—this attention doesn't grant the world its importance. It does not imbue the grass or the trees or the people with the presence of God. Rather, it is one practice of recognizing the life, the createdness, and the holy in and throughout creation. It is giving the hum of God's presence, the life and love that is within everything spoken into being, my time. Acknowledging it with my mind. My heart. My full being.

There is no danger, in my mind, in extending the same love and generosity that I do to my human neighbours towards the rest of the world. My love of the ecological world pulls me into a rejoicing, a more fully human way of being. As I come to recognize my surroundings, my ecosystem, I come to recognize myself as an earthly being that takes part in this ecosystem. Breath in my lungs and soil beneath my feet—these connect me to all of creation. Water. Apples. Toothpaste. Morning coffee. Afternoon walks with friends. Driving to work. Everything I do is connected to something else. As I live this lavishing of attention, I understand more and more my place in it all; layer by layer, leaf by tree by river by stream by rivulet of water trickling down the sidewalk to the drain, another moment in my understanding of my role in the world.

One of my favourite ecological facts is not actually about the study of ecology at all. It is about the etymology of the word. The root of "ecology"—eco—is derived from the Greek word "oikos" meaning "household" (Pimm, Smith & Leo, 2022). A house. A home. The study of ecology is the study and research of the relationships between all living things and their surroundings: relationships within a household. Not just individual species or specific climates or forest plots. Ecology requires acknowledging a lack of borders, the ephemeral nature of boundaries, and the truth that the world is overwhelmingly interconnected and conjoined. As we keep this in mind, ecosystems transform from discrete biomes defined by plant communities, or climate, or divides drawn on human maps, and become networks sustaining members that are *related* to each other. At the time I first learned this fact, the importance of connection was not new to me. I was fully aware of the importance of community, how much I craved social interaction, how I valued my slowly growing circle of friends. I knew that there was a world outside of myself that I relied on for food and water. I had even begun to take the bus to the farmers' market to spend my meagre paycheque on corn and carrots and sweet fall apples. It was a joy and privilege to meet the farmers who grew these foods, and I revelled in my ability to support local agriculture, connecting to this world outside the city. But this? This was an entirely new and profound level of connection. The use of such a human word—household—to describe something that I had ultimately believed to be different from myself changed my understanding of the world. I wasn't as separate as I had first believed. I was—am—in community with the entirety of my ecosystem; the rest of the planet; the whole world, whether I am aware of it or not. I am, as my Opa used to say, "a citizen of earth!" A part of the household of all things.

In so many ways, our identity as members of an ecosystem has been eroded. Even coming from a family that values farming—grandparents who grew much of their own food, and parents who are slowly transforming their suburban backyard into a lush garden—it is easy to forget that ecosystems are not separate from humanity. Years of urbanization and modernization have slowly changed the landscapes and seemingly cut us off from "nature" as many people think of it. Like the treeless fields losing soil inch by inch, urbanization has washed away ways of life that explicitly connect us to our location. Familiarity with our surroundings, awareness of where our food comes from, who brings our water and where it goes when it drains from the bathtub or sink: these are all facts that are easy to ignore, forget, or never learn in the first place. And the result is that we don't even realize how much we depend on—are in relationship with—the rest of the earth. Instead, the majority of us are urban citizens, admonished by many churchgoers to reject the world. We are disconnected from fields and streams, seeing them as separate and needing of preservation. They are acknowledged as beautiful and worthy of care and attention, sometimes even the subject of hymns or homilies. But they are not seen a part of us. We need to know that caring for the polluted river is another way to care for the "least of these." That adoration for the mountains and the weeds is adoration not only for the Creator, but for the women who have lived on and throughout the mountains for centuries, and for the men who pull weeds from soil for a living. We are part of a much larger household, one that includes every creature in our neighbourhood, every farmer in every country growing oranges and winter lettuce, every technician at the water treatment plant, every minute nematode slowly breaking down the leaves that fall from the linden trees lining the roadsides, every pocket of lichen slowly transforming light into life. There are plants, animals, micro-climates, small and large interactions forming and reshaping the world. There are webs of connection, vibrancy, and life everywhere you look. In any place that you look. Once you begin to uncover these relationships in ecology, the threads connecting all of creation are revealed. Like a shimmering tapestry—they spread out across the planet, uniting each living thing to the other.

The idea of a household, of ecosystems being based on relationships, transforms the idea of community from a fact discussed and debated into a way of life. There is so much world beyond our daily experiences, so much of the earth that we rely on and don't even realize it. The earth is revealed to be integral, intimately related to every moment. Even my breath, even just breathing, is only possible because of the existence of a thousand others: trees, grasses, lichens, algae, who all continuously transform my carbon dioxide to oxygen. Just by breathing we are in constant communion with the world, infused with the breath and words of The Creator. Through this realization of joint participation in our ecosystem, there is a profound connection to those beyond ourselves and beyond our species. It is like tuning in to the voice of creation—something that is usually white noise in the background becomes unmistakable once you realize that it is there. This integration into the fabric of creation is a "oneness" akin to the belonging constantly reflected and repeated in scripture. A relationship formed of mutual belonging. As Christians, we are often so quick to reject the earthly, that we forget God created us with the same breath that was used to speak the earth, this entire diverse and wonderful earth, into being. And God called it good! We are so consistent in reminding ourselves that we do not live by bread alone, that we miss the significance of needing bread, of being sustained in our physical bodies by the world that God created. This is the gospel. The good news of belonging and generosity and love. This is a call towards connection. Towards "a spirituality for this world that [allows] you to delight in this world and not just to seek to escape it, to get away from it, but to find God in it and through it" (Rohr, 2020). Towards a deeper knowing of God.

I worship a God that has always been, and will always be, relational. At the beginning of all things, God was Three, moving in unison, speaking together over the waters, breathing life into the creatures of the air and sea and earth. Creating together. In the beginning, there was connection. Relationship. I follow a man who celebrated community, a man who moved in radical ways against division. And I choose to participate in a community without borders, recognizing the vast world that I am connected to. The love of God and the fellowship of the Holy Spirit ranges far beyond the four walls of any building. The community of love and care and generosity made clear in the Gospels is only made more tangible and liveable through the ecosystems that we all participate in. Our earthiness is not a taint, rather it is the foundation from which we build the Kingdom of God. Because of this, environmental ethics is not a separate branch of theology. Environmental ethics becomes a daily practice, embodied, rather than merely ideas. We are able to enter into stewardship, live and breathe it through our very existence. Rather than becoming mired in dogmatic thinking, rigid rule-following, or heavy-handed judgement, we are able to move forward into more sustainable, more ethical, and more connected ways of being. Focusing on love and attention, letting this awareness of connection guide the unveiling of creation is the foundation to environmental responsibility. It gets at the heart of what I believe: moving away from division and hurt and perfectionism, towards love and relationship. Towards the God of love and relationship.

There are wind maps on the water's surface. All around you. Often, it is something you can neither see nor feel. But if you slow yourself, take a breath, and tune in to the rhythms of creation, you will see traces of these movements and interactions beyond yourself. Open up your mind and eyes and ears and nose and skin to the world and see that the world is good! Take a moment to look up and find the moments of wonder, of wind and water on a long drive. Allow yourself to well up with wonder at the ecosystem you are a part of, whether it is outside in your local park or forest, or in the grocery store admiring the vast diversity of fruits that we have access to in December. Move steadily and daily towards delight in the world and, through this, find God. Come awake to the aliveness of the world, the holiness of creation. Remember that you are a part of something bigger. Allow the household of all things to shape your ethic, with loving relationship as your foundation. Open yourself to awe and wonder. Begin, moment by moment, to learn the language of beauty and belonging. Find the threads of connection weaving you into relationship with the world, and fall into the wonderful intimacy of that community, a household with everything else in existence. Allow this to be your ethic, an ethic of care and connection. Bask in simple movements, be made aware of wind that you cannot feel, and be reminded of the world that exists beyond yourself. This is the practice of turning towards beauty: the practice of lavishing attention.

References

American Heritage Dictionary of the English Language. 5th ed. New York: Houghton Mifflin Harcourt Publishing, 2016.

Coleman, Daniel. *Yardwork: A Biography of an Urban Place*. Hamilton, ON: James Street North Books, 2017.

Egerton, Frank N. "History of Ecological Sciences, Part 47: Ernst Haeckel's Ecology." *Bulletin of the Ecological Society of America* 94, no. 3 (2013): 222–244.

Francis. *On Care for Our Common Home*. Toronto: Pauline Books and Media, 2015.

Pétrement, Simone. *Simone Weil: A Life*. New York: Pantheon Books, 1976.

Pimm, S. L. and Robert Leo Smith. "Ecology." *Encyclopedia Britannica*, 26 December 2022. www.britannica.com/science/ecology.

Rohr, Richard. Interview with Brie Stoner and Paul Swanson, *Another Name for Everything*, Podcast Audio, 4 April 2020. https://cac.org/podcasts/environmental-awareness-rooted-in-franciscan-spirituality/

3 An Ecocritical Reading of *Jesus of the Deep Forest*

Susan Emily VanZanten

During the twenty-first century, ecocriticism has gone global, with postcolonial ecocriticism's attention to texts from the global south employing a sociocentric rather than an individualistic lens, often with a concern for environmental justice (Buell, "Ecocriticism," 2011, 87–115).[1] Although the so-called White thesis argues that Christianity is principally responsible for today's environmental problems because of its affirmation of human dominion over the natural world, the literature of the global south contains a rich collection of environmental writing often in dialogue with Christian theology and practice (White, "Historical Roots," 1967, 1203–1207; Burbery, "Ecocriticism and Christian Literary Scholarship," 2012, 189–214). Postcolonial ecocriticism to this point has primarily focused on East Asian texts, but study of African texts is increasing.[2] Some African texts, we are discovering, problematize the White thesis by endorsing permeable boundaries between divine transcendence and the natural world as an essential part of an African Christian belief. Instead of dominion over nature, African Christians often affirm a loving interdependence with nature. They prize community over commodification.

This essay examines what happens when a postcolonial African text concerned with both the environment and Christianity, *Jesus of the Deep Forest*, is studied with an ecocritical eye and ear. Colonial African texts habitually erased the traces of Indigenous people and the effects of colonial conquest and rule; Nixon notes that colonized people "have been repeatedly naturalized as an object of heritage to be owned, preserved, and patronized rather than as a subject of their own land and legacies" (Nixon, "Ecocriticism," 2005, 25). However, *Jesus of the Deep Forest* presents the (mediated) voice of an Indigenous African speaking by means of a traditional genre about the significance of physical place in Christian terms. Exemplifying the explosive growth in Christianity in Africa since decolonization, *Jesus of the Deep Forest* does not cast Africans as victims requiring Western protection or proselytization but rather highlights the role of African agency in the spread of Christianity through Indigenous churches. The poem is an expression of what African religious historian Lamin Sanneh calls "World Christianity"—a movement in which Indigenous people discover Christianity on their own terms rather than capitulate to European Christianity (Sanneh, "Whose Religion?," 2003). This discovery process, says missiologist Andrew Walls, *converts* rather than *proselytizes*.

DOI: 10.4324/9781003366744-3

Assuming that cultures have fixed boundaries, the proselyte entirely discards former beliefs and practices in favour of new ones, but the convert acknowledges the truth of many Indigenous beliefs and practices, transforming the old into something new by redirecting the material, literary, and social practices of a local culture towards Christianity (Walls, "Some Signposts," 1997, 148).

Jesus of the Deep Forest is a remarkable and often overlooked work of African oral literature composed and performed in the 1970s by Afua Kuma, a non-literate Ghanaian village woman of Akan heritage. According to *The Concise Oxford Dictionary of the Christian Church*, Christian activity in Ghana commenced in the fifteenth century, but extensive missionary work only dates from the early nineteenth century. During the First World War, European missionaries were expelled from the country, but local Ghanaian leadership significantly grew church membership: "Preachers without missionary tutelage caused large movements towards Christianity. Where Christianity had already been established, Churches stressing prophecy and healing developed, incorporating elements of traditional culture" ("Ghana," *Concise Oxford Dictionary of the Christian Church*, ed. Livingstone, 2013). Ghana achieved independence from Great Britain in 1957, and by the late 1970s, when Kuma composed the poem, almost 53% of Ghanaians identified themselves as Christian (Pobee, "Religion and Politics," 1987, 44). From the 1980s on, an "explosive growth" in Christianity in Ghana took place, especially in Independent Pentecostal or Charismatic churches. Today more than 72% of Ghana's population is Christian ("Ghana," *World Christian Encyclopedia*, ed. Johnson and Zurlo, 2019).

Afua Kuma (1900–1987) was brought up as a Christian by Presbyterian parents.[3] She and her husband raised nine children on their small family farm in the Ashanti highlands of Ghana, where she also served as a midwife. When two of her sons joined the Church of Pentecost, an African Independent Church that is the single largest Protestant denomination in Ghana ("Ghana," *World Christian Encyclopedia*, 2019), Kuma left the Presbyterian Church to join them. At the age of 70, Madam Kuma, as she was respectfully termed, heard a sermon calling Christians to give continual praise to God, followed by an invitation to pray. In response, much to the congregation's surprise, she spontaneously composed and performed *Jesus of the Deep Forest*. When she was later asked to perform the prayer again, she recited it in exactly the same sequence as before. Consequently, for the last 17 years of her life, Madam Kuma travelled throughout Ghana performing her poem at Church of Pentecost retreats, conventions, Easter rallies, and evangelistic crusades. Ghanaian linguist Akosua Anyidoho comments, "The old lady had a phenomenal memory if we consider the fact that she could stand among a group of worshippers and pray for a long time (as a study of different recordings shows)" (Anyidoho, "Techniques," 83). *Jesus of the Deep Forest* was first recorded in the late 1970s and then transcribed in the Akan (commonly known as Twi) language of its performance. The poem was then translated into English by a Catholic priest, Father Jon Kirby, a respected Akan ethnographer. In 1981 the poem was published in Accra in two editions: Twi and English. The Twi version, *Kwaebirentuw ase Jesu—Afua Kuma ayeyi ne mpaeb*, has regularly sold out in Ghana, but the English translation has been less popular (Bediako, *Jesus and the Gospel*, 2004, 14). The printed poem is long, with 185 stanzas extending over 46 pages. Within her cultural context, Madam Kuma is highly celebrated; Anyidoho calls her "one of the most powerful creative verbal artists that Akan society has known in contemporary times" (Anyidoho, "Techniques," 73). Similarly, *The Cambridge Companion to Feminist Theology* cites Madam Kuma's work as an exemplary paradigm for African women theologians (Oduyoye, "Jesus Christ," *Cambridge Companion*, 2002, 151–170).

Generically, *Jesus of the Deep Forest* is an *apae*, a highly respected and intricately structured Ghanaian form of praise poetry. The genre of this work as oral performance is crucial to consider from a Christian ecocritical perspective, as I will discuss later. Many African cultures include both informal praise singing (as a mother to her child) and a more formal, public poetry performed by an official singer to celebrate a king or chief. Oral poets such as *griots* of West Africa or the *imbongi* of South Africa are still central figures at public ceremonies today. In 1996, during a visit to the University of Natal, I witnessed a long exuberant praise song for then-President Nelson Mandela performed by the University's Zulu student praise singer, and many people around the globe heard Xhosa praise singer Zolani Mkiva's performance at Mandela's funeral in 2013 on BBC radio. In Ghana, the praise poet typically is male and is called *Okyeame*, "the royal spokesman" (Anyidoho, "Linguistic Parallels," 1991, 67). Madam Kuma thus assumed the role of the royal spokesperson for Jesus, as a genre typically performed by a male *Okyeame* to praise a chief is redirected by a female poet to celebrate Jesus as the ultimate ruler. The principal topics of traditional *apae* are the military exploits of the king or chief, his generosity towards his people, and his superior hunting skills. Formally, the *apae* employs an elaborate structure of repetition and parallelism rather than narrative, rhythm, or rhyme. Printed stanzas have varying numbers of lines, each with varying numbers of words; there is no set pattern. The printed line divisions are most likely determined by the original delivery, in which a group of words would be pronounced with one breath, followed by a pause. The stanza endings of a performed *apae* are typically indicated by lowering the voice (Yankah, "To Praise or Not to Praise," 1983, 390). The *apae*'s primary formal unit is the praise name, a laudatory title that declares the celebrant's notable qualities and deeds. Praise names are then supplemented, expanded, and repeated in subsequent phrases, clauses, lines, and stanzas. An *apae* contains three kinds of praise names: personal names, traditional "strong appellations," and metaphoric names (Yankah, 385). In *Jesus of the Deep Forest*, the personal name *Jesus* appears over a hundred times, followed by a clause signifying an action, character trait, or metaphoric name, such as "Jesus, the Seer among prophets/who always speaks the truth," or "Jesus, you are the Hunter!"[4] In effect, the *apae* presents numerous snapshots that cumulatively produce a complex, multifaceted, mutable image of the subject of praise rather than recounting their life events in chronological order, as a Western hagiography might. *Jesus of the Deep Forest* is an extended song of praise in honour of Jesus as the ultimate ruler within the context of a modernizing African society in which the environment plays a key role.

The poem presents the perspective of a member of the impoverished global south on the relationship of humans and their environment that is quite unlike that found in other African texts. Carminero-Santangelo and Myers note that many colonial and postcolonial texts such as travel writing, hunting narratives, and self-discovery memoirs

> depict African environments as places of wilderness in which nature is identical with, absent of, or threatened by Africans themselves. This vision reinforces binaries of nature and culture, wilderness and civilization, and human and non-human that encourage an instrumentalist approach to both places and people.
> (Carminero-Santangelo and Myers, "Environment," 14)

The romantic tendencies of ecocentrism thus write over the Indigenous other. *Jesus of the Deep Forest* refuses such binaries, objectification, and instrumentalist thinking. Some African novelists and poets, such as Ken Saro-Wiwa, Ben Okri, Jake Mda, and Imbolo Mbue, have written about the serious negative impact of colonial capitalism and postcolonial global corporations in Africa, examining how deforestation and oil extraction have transformed and exploited both the natural environment and local communities. But Kuma's approach is different, in terms of both the genre she employs and the way she depicts the environment; she celebrates the Ghanaian Akan environment through an "aesthetic of proximity" (Iheka, "Naturalizing Africa," 2018, 21) focused on the interconnection of human lives with others in the natural environment, both physical and spiritual. That interconnection is understood in Christian terms.

As both the title and cover photograph of *Jesus of the Deep Forest* suggest, the forest is the central physical environment of the poem. The title opens with the primary praise name *Jesus* followed by a prepositional phrase that locates him physically, culturally, and spiritually: "of the deep forest." The cover of the English edition features a black-and-white photograph of a dirt road leading into a woodland of tall trees and thick undergrowth, inviting the reader to journey into the deep forest. Ghana's thickly canopied forests are much more than landscapes; they are viewed as sacred sites inhabited by gods and demons, and traditional cultural practices protect forests by regulating usage to prevent deforestation, contamination of watersheds, and destruction of flora and fauna. The forest was not a forbidden place, with a few exceptions, but rather one that required careful and respectful use. Traditional Akan values hold that communities will survive only if forests and other natural resources endure and that humans were created to dwell in close relationship with other creatures (Asante, Babio, and Boadu, "Indigenous Cultural Practices," 2017). Fruit, bushmeat, firewood, and lumber are gifts from the forest to be praised, not exploited. The Akan life of *Jesus of the Deep Forest* embodies, within a different sociocultural location, what American conservationist Aldo Leopold called "a land ethic": "A land ethic changes the role of *Homo sapiens* from conqueror of the land-community to plain member and citizen of it. It implies respect for his [or her] fellow-members, and also respect for the community as such" (Leopold, *Sand County Almanac*, 1987, 204).

The deep forest is a crucial part of the Akan community, and the "signs and wonders" performed by Jesus and celebrated throughout the poem are often associated with the forest. As a generous miracle worker, Jesus transforms barren land into fertile forest: "O Master of signs and wonders: you stretch your hands/over the deserts, and a forest appears!" (9). Jesus *creates* the forest, but he also *is* the forest. The "many titles" of Jesus that the poem celebrates include that of the deep forest: Jesus is "the great forest canopy that gives cool shade:/the Big Tree which lifts its vines/to peep at the heavens,/the magnificent Tree whose dripping leaves/encourage the Luxuriant growth below" (5). Later the poem directly addresses Jesus: "You are the deep forest which gives us tasty foods./The forest of cane and thorns,/where wander *kente* weavers,/in search of shuttle and loom" (37). As the forest, Jesus provides protection from the sun, flourishing fertility, delicious food, and material to construct the looms on which costly *kente* cloth of silk and cotton is woven. The forest praise name celebrates lavish provision, not bare bones survival. Within this praise name, Jesus is both the subject and the object, the transcendent creator and immanent incarnation. The cultural centrality of the forest for Akan life, both physical and spiritual, is captured here.

The forest's provision is also acclaimed through Jesus's hunting prowess: "He sets a trap in the forest and catches fish" (6), and "You [Jesus] stand in the sea while you hunt in the forest" (9). The poem depicts Jesus hunting buffalo, crocodiles, birds, fish, crabs, and—most significantly and repeatedly—the elephant—considered "the king of all animals" in Ghana. Jesus is "***Oserekyi Sakyi*****:** the Elephant Hunter,/whose family's cooking pots/have no place for little mushrooms" (27). This passage opens with a traditional Akan praise name (some 50 of which occur in the poem), a strong appellation made up of multiple Twi complex nouns presented in bold type and usually followed by an English translation. The retention of the Akan praise names in the English translation indicates concepts that can't be fully translated into English, aspects of Jesus that can only be named in Twi, as Akan culture opens up new ways of perceiving, understanding, proclaiming, and praising Jesus. Most Western readers cannot fully comprehend what it means that Jesus is the elephant hunter. Jesus hunts elephants in miraculous ways, demonstrating his immense power: "Jesus walks away on air, his back to us;/when he turns around and comes back,/he is holding an elephant" (35). Although rifles and bullets are available, Jesus uproots a kapok, a gigantic tree of the tropical forest canopy, to kill another elephant (19). He also assists human hunters, helping them to remove the elephant from a pit dug to trap him (33). Even more striking is the praise name that designates Jesus as the very gun that kills the elephant: "Among powerful rifles, you are the Elephant-Gun./Jesus enables the hunter to kill the elephant" (9). In a swiftly moving stream of praise names and phrases, Jesus is the ultimate elephant hunter, the helpful ally of human hunters, and the lethal elephant gun, but Jesus is also the elephant itself: "The great Elephant of the forest,/who knows its secrets,/and walks its arbors, without hindrance" (33). Just as Jesus creates the forest, provides gifts from the forest, and is the forest, so he permeates the image of the elephant, assuming different roles in different scenarios. In traditional Ghanaian thought, the highly respected elephant is understood to be a valuable but dangerous physical creature and a powerful malevolent spirit that can survive after physical death (Kwarteng, "Elephant in Pre-Colonial Ghana," 2006, 9). But Jesus defeats such evil manifestations: "He, the sharpest of all great swords,/has made the forest safe for the hunters./The *mmoatia* he has cut to pieces;/he has caught *sasabonsam*/and twisted off its head" (19). In Akan thought, the *mmoatia* are dwarf-like troublemakers, while the *sasabonsam* is variously described as the forest monster, Satan, or, sometimes, the elephant (Koranteng-Green, "Multiplicity," 2018, 103–104). The poem brings together this emblem of powerful evil with the hunter praise name:

> He is the Hunter gone to the deep forest.
> *Sasabonsam*, the evil spirit,
> has troubled hunters for many years.
> They ran in fear,
> leaving their guns behind.
> Jesus has found these same guns,
> and brought them to the hunters
> to go and kill the elephant.
>
> (19)

Jesus, as the elephant, replaces the evil spirit, providing both material and spiritual assistance for human hunters.

After personal names and strong appellations, the third type of *apae* praise name is usually identified in Western literary criticism as metaphoric, such as the forest, elephant, gun, and sword. But considered as metaphors, such praise names appear inconsistent and contradictory. The personal name of Jesus, the traditional Twi appellations, and the metaphoric designations are all complexly entwined and frequently shift their referent. *Jesus of the Deep Forest* uses numerous animal names beyond *elephant* for Jesus, including python, mamba, crab, lamb, whale, ram, lion, and leopard, and Jesus is also the hunter of many of these same animals. Vegetative, geographical, cosmological, and meteorological titles occur even more frequently; Jesus is a tree, vine, beans, "the best of yams," (32), "glistening Water-lily" (13), rock, lightning, mountain, wind, "the fiercest of storms" (9), hurricane, rain, sunshine, moon, breeze, river, and morning light. He is ever-present in the poem's natural environment. But he is equally present in the sociocultural environment. Moving away from more conventional metaphors, *Jesus of the Deep Forest* calls Jesus an iron rod, weaving loom, boat, pestle, hoe-handle, string of beads, grass hut, grinding stone, sword, house, farm, lantern, city, Caterpillar (bulldozer), crossbeam, and elephant gun. Jesus also assumes many different social roles beyond the traditional king, chief, or military hero: he is a magician, porter, farmer, teacher, steward, lawyer, chief of police, doctor, sword-carrier, counsellor, political arbitrator, linguist, driver, mother, friend, swimmer, and mechanic. The poem locates Jesus as an inherent part of the contemporary African physical, social, and cultural environment, in both nonhuman and human worlds. The physical environmental elements are not allegories or symbols, metaphors or metonyms, but rather shifting cultural markers potentially full of immanence for those with eyes to see or ears to hear. Western literary concepts are not adequate to describe the rapidly shifting conjunctions of praise names, phrases, and stanzas. As Caleb Monday Jonah, from the University of Wakari, Nigeria, explains, "An understanding of a former colonized people and their attachment to place can be understood when binaries are dissolved" (Jonah, "Critical Analysis," 2020, 37).

The physical environments and attachment to place depicted in the poem extend beyond the deep forest; the second-most important site is agricultural land. Like the Akan people, Jesus is both Hunter and Farmer, "Chief of farmers;/ who give us our produce!" (28, 38). The theme of agricultural abundance echoes biblical stories of miraculous feedings: "You take a single grain of corn, grind it, roast it,/and then go and plant it and, look! Now the grain has borne fruit, and filled two hundred bags,/with some left over!" (15). As a farmer and supporter of farmers, Jesus engages in activities that Western environmentalists may find questionable; he "light[s] a bush-fire to clear a farm" (15); He is "the Caterpillar who plough up the land/for the Northerners to plant their guinea-corn" (37); and he gives a "cutlass . . . to your workers,/to cut down the forest" (38). The poem's inclusion (and celebration) of technology along with its praise of forest clearing and elephant hunting introduces an element of what Karen Thornber terms "eco ambiguity": portrayals of inconsistent human attitudes, behaviours, and information concerning the environment, when considered from a Western perspective (Thornber, *Ecoambiguity*, 2012). *Jesus of the Deep Forest* does not call for forest preservation or rejection of Western technology, but its profound affirmation of forest and farm as holy presence, gift, and community reminds readers of a voice that should be honoured in the midst of competing demands in the global south for agricultural progress and environmental sustainability, human sustenance, and animal preservation.

Other physical places in *Jesus of the Deep Forest* include towns, villages, the savannah, the sea, streams, and a city. However, this is not the crowded streets of Accra or any other temporal city; this is a "holy city":

> Let's go to Jesus' City,
> Called *Soe-di-bi*
> (Put down your load and have something to eat).
> He is the great city
> Whose children are always well fed.
>
> (13, 37)

The city, like the forest, is associated with sustenance, and its Twi name, as translated in the book's glossary, means "take and eat," the sacramental invitation for Christian communion or the Lord's Supper. Jesus provides all the necessary food, from the "tasty food" of the forest, to the cultivated yam and corn, to the fish of the sea and streams, to the life-giving miracle of his body and blood. The descriptions of the holy city are clearly taken from the biblical book of Revelation, with imagery of gold, everlasting light, precious stones, angels, and a grand ceremonial march celebrated with song and dance, a triumphal progress like that of the traditional chief. In this procession, the natural world joins with the *Okyeame* and other humans in singing praises:

> There is a wild noise,
> like that of rushing streams.
> Yet it is not the sound of streams we hear;
> it is the thunder and lightning serenading Jesus!
>
> The wind clutches a golden banner,
> its left arm laden with beads and bracelets of gold,
> as it dances before Jesus' face.
>
> (13)

While Western readers may see personification in such passages, from an African world view, they reflect the common agency of the natural and human worlds to worship and celebrate Jesus. As Egya says:

> . . . the natural world . . . does not imply the same thing to sub-Saharan Africans as it does for non-Africans, especially those in the so-called developed societies. . . . Africans, whether in the precolonial time or in the present state of struggling modernity, do not see nature as the *other*, against which to define themselves . . . Africans, especially those born and raised on the continent, understand the connections between physical and spiritual (seen and unseen) entities. They tend to see themselves as part of nature.
>
> (Egya, "Out of Africa," 2020, 67, 68)

The Hebrew Testament reflects a similar perspective in Isaiah, when the prophet predicts, "You will go out in joy and be led forth in peace; the mountains and hills will burst into song before you, and all the trees of the field will clap their hands" (Isaiah 55:12, New International Version). In other joyful processions recounted in the poem, all the human members of the community participate—from the "chiefs, so majestically arrayed/in beauty, grandeur and crowns of gold," to "young men [who] rush to gather,/dancing and making him welcome," to the women who "dance and play,/encouraging all with shouts of praise/before the face of Jesus" (21). Jesus has absorbed the traditional role of the Akan chief, who then becomes another member of the environment—physical, social, and spiritual—praising Jesus. The entire community, both human and nonhuman, partakes in a flourishing wholeness, bounty, and joy.

The poem's complex ecology defying Western binaries is also apparent in its status as performance. The praise singer speaks not for herself as an individual but for the community as a whole. Instead of a personal epiphany such as that experienced by American transcendentalists or British Romanticists, we witness a communal ecstasy. *Jesus of the Deep Forest* is not a private literary production but a public event, involving everyone who hears and witnesses it; it is not a lyric but an *apea*. In performance, Madam Kuma directly addressed a crowd of people, and much drama would have occurred through fluid, improvised reactions, and exchanges. In a transcribed *apea*, such effects are created through shifts in person and address. The voice seldom uses a lyrical "I," usually speaking in a communal "we." Even when the "I" appears, it is most often situated in community: "You and I" (14, 27); "I and my friends" (29). However, most of the poem is addressed to Jesus in plural third person, *you*, which makes the occasional shifts in person and address especially notable. At several points, the praise singer turns from speaking to Jesus to addressing the audience: "If Jesus should call you, smile/he gives gifts freely" (23). An evangelistic altar call appears near the end of the poem: "My brother, you who are standing there,/ come to Jesus" (44). We might imagine Madam Kuma speaking directly to someone standing in the audience as she utters this invitation. Several times she also addresses one specific group of people—the *osofo*, the Christian pastors, admonishing them to take up their responsibilities: "Osofo . . . God has made of you a guardian" (30).

In considering *Jesus of the Deep Forest*, we must acknowledge the limits of analysis of orature. When an oral performance is transcribed and translated, we miss "the full actualization of the poem as an aesthetic experience" (Finnegan, *Oral Literature*, 2012). We cannot experience the richness of performance: gestures, facial expressions, voices, tones (especially significant in African languages), rhythms, and musical accompaniment. Furthermore, we miss the social context and exchanges with the audience. John Foley introduced the term "oral-derived texts" for works that translate performance into writing and fix fluid orature into a stable form. In *How to Read an Oral Poem*, he calls such texts "Voices from the Past" and concedes that "direct experience of that oral poetry lies forever beyond our reach" (Foley, *Oral Poem*, 2002, 45). Nonetheless, while a live performance allows one kind of interpretation, analysis, and appreciation, an oral-derived text allows another. Without experiencing the aural dynamics and social context of a performance, we can still read, study, and discuss the composition as delivered, recorded, transcribed, translated, and printed. However, in doing so, we must clearly admit that, as Bradbury says, "What survives is an echo of an oral performance" (Bradbury, "Traditional Referentialities," 1998, 136). Madam Kuma's voice is mediated not only through language and culture but also through the stability of a written text, which hints at the uncertainties and obfuscations involved, from the untranslatable Twi appellations and words, to the unstable metaphors, to the lack of a cohesive narrative, to the final African rejection of simple Western binaries of human and nonhuman, of body and spirit, of God and creation.

In "Out of Africa: Ecocriticism beyond the Boundary of Environmental Justice," Sule Emmanuel Egya argues that African "ecocriticism should recognize the commonalities between the natural and human worlds that have been understood to exist in Africa." He continues:

> The emphasis on environmental justice has suppressed a crucial dimension of environmental literature in Africa, by failing to foreground natural worlds and their role in shaping thoughts and ways of living, as well as in the preservation and conservation of diverse life-forms.
>
> (Egya, "Out of Africa," 2020, 67)

While Egya argues that critics should acknowledge African spiritualities, the belief in re-incarnation, and the agency of nonhuman beings, this essay takes a different approach: recognizing the African commonalities of the natural and human world and their close spiritual and geographical kinship but identifying the way in which those concepts and a traditional creative form have been redirected to African Christianity. Hearing the (admittedly mediated) voice of an Indigenous African woman who has intimate visceral knowledge of the Akan environment of forest, farm, and village, we may learn new ways of interacting with and thinking about the nonhuman world, as well as new ways to support the local sustainability of Indigenous agricultural communities.

Notes

1 For other accounts of postcolonial ecocriticism, see Caminero-Santangelo and Myers, *Environment at the Margins: Literary and Environmental Studies in Africa*, 2011; Cilano and DeLoughrey, "Against Authenticity," 2007; Huggan and Tiffin, *Postcolonial Ecocriticism* 2010; Nixon, "Environmentalism and Postcolonialism," 2005; and Vital and Erney, eds., *Postcolonial Studies and Ecocriticism*, 2006–7.

2 Caminero-Santangelo and Myers, *Environment at the Margins*, is the first book-length discussion of African literature from a postcolonial ecocritical perspective. Journal essays include Caminero-Santangelo, "Different Shades of Green," 2007; Egya, "Out of Africa," 2020; Iheka, *Naturalizing Africa*, 2018; Jonah, "Issues in Postcolonial Criticism," 2020; Slaymaker, "Ecoing the Other(s)," 2001; and Vital, "Toward an African Ecocriticism," 2008.

3 Portions of this and the following paragraph are adapted from VanZanten, "World Christianity," 2002. Biographical information comes from Kirby, "Preface," to *Jesus of the Deep Forest*, 1981; and Anyidoho, "Techniques of Akan Praise Poetry," 2000.

4 Kuma, *Jesus of the Deep Forest: Prayers and Praises of Afua Kuma*, trans. Jon Kirby (Accra, Ghana: Assempa Press, 1981), 29, 33. The English version of the poem does not include line numbers, so subsequent in-text citations will be to page numbers.

Bibliography

Anyidoho, Akosua. "Linguistic Parallels in Traditional Akan Appellation Poetry." *Research in African Literatures* 22, no. 1 (1991): 67–81.

———. "Techniques of Akan Praise Poetry in Christian Worship: Madam Afua Kuma." In *Multiculturalism and Hybridity in African Literature*, edited by Hal Wylie and Bernth Lindfors, 71–86. Lawrenceville, NJ: African World Press, 2000.

Asante, Eric Appau, Stephen Ababio and Kwadwo Boadu. "The Use of Indigenous Cultural Practices by the Ashantis for the Conservation of Forests in Ghana." *SAGE Open* 7, no. 1 (2017). https://doi.org/10.1177/2158244016687611.

Bediako, Kwame. *Jesus and the Gospel in Africa: History and Experience*. Maryknoll, NY: Orbis Books, 2004.

Bradbury, Nancy Mason. "Traditional Referentiality: The Aesthetic Power of Oral Traditional Structures." In *Teaching Oral Traditions*, edited by John Miles Foley, 136–145. New York: MLA, 1998.

Buell, Lawrence. "Ecocriticism: Some Emerging Trends." *Qui Parle* 19, no. 2 (Spring/Summer 2011): 87–115.

Burbery, Timothy J. "Ecocriticism and Christian Literary Scholarship." *Christianity and Literature* 61, no. 2 (Winter 2012): 189–214.

Caminero-Santangelo, Byron. "Different Shades of Green: Ecocriticism and African Literature." In *African Literature: An Anthology of Criticism and Theory*, edited by Tejumola Olaniyan and Ato Quayson, 688–707. Oxford: Blackwell, 2007.

——— and Garth Myers, eds. *Environment at the Margins: Literary and Environmental Studies in Africa*. Columbus: Ohio University Press, 2011.

Cilan, Cara and Elizabeth DeLoughrey. "Against Authenticity: Global Knowledges and Postcolonial Ecocriticism." *ISLE: Interdisciplinary Studies in Literature and Environment* 14, no. 1 (2007): 71–87.

Egya, Sule Emmanuel. "Out of Africa: Ecocriticism beyond the Boundary of Environmental Justice." *Ecozon@* 11, no. 2 (2020). https://doi.org/10.37536/ECOZONA.2020.11.2.3495.

Finnegan, Ruth. *Oral Literature in Africa*. Cambridge: Open Book Publishers, 2012. https://doi.org/10.11647/OBP.0025.

Foley, John Miles. *How to Read an Oral Poem*. Champaign, IL: University of Illinois Press, 2002.

Huggan, Graham and Helen Tiffin. *Postcolonial Ecocriticism: Literature, Animals, Environment*. London: Routledge, 2010.

Iheka, Cajetan. *Naturalizing Africa: Ecological Violence, Agency, and Postcolonial Resistance in African Literature*. Cambridge: Cambridge University Press, 2018.

Johnson, Todd and Gina A. Zurlo, eds. *World Christian Encyclopedia*. Edinburgh: Edinburgh University Press, 2019.

Jonah, Caleb Monday. "A Critical Analysis of Issues in Postcolonial Criticism." *Ibadan Journal of Humanistic Studies* 30 (2020): 37–51. https://www.ajol.info/index.php/ibjhs/article/view/208962.

Kirby, Jon. "Preface." In *Jesus of the Deep Forest: Prayers and Praises of Afua Kuma*, vol. 3. Accra, Ghana: Assempa Press, 1981.

Koranteng-Green, E. M. Kwapong. "The Multiplicity of Ahonhom (Spirits) in the Akan Spiritual Cosmology." *Journal of Philosophy, Culture and Religion* 39 (2018): 101–106. https://iiste.org/Journals/index.php/JPCR/article/view/45855.

Kuma, Afua. *Jesus of the Deep Forest: Prayers and Praises of Afua Kuma*. Translated by Jon Kirby. Accra, Ghana: Assempa Press, 1981.

Kwarteng, Kwame Osei. "The Elephant in Pre-Colonial Ghana: Cultural and Economic Use Values." *Journal of Philosophy and Culture* 3, no. 2 (June 2006): 1–32. https://doi.org/10.4314/jpc.v3i2.71860.

Leopold, Aldo. *A Sand County Almanac, and Sketches Here and There*. Oxford: Oxford University Press, 1987.

Livingstone, Elizabeth A., ed. *Concise Oxford Dictionary of the Christian Church*. Oxford: Oxford University Press, 2013. eISBN: 9780191727214.

Nixon, Rob. "Environmentalism and Postcolonialism." In *Postcolonial Studies and Beyond*, edited by Ania Loomba, Suvir Kaul, Matti Bunzl, Antoinette Burton and Jed Esty, 233–251. Durham, NC: Duke University Press, 2005.

Oduyoye, Mercy Amba. "Jesus Christ." In *The Cambridge Companion to Feminist Theology*, edited by Susan Frank Parsons, 151–170. Cambridge: Cambridge University Press, 2002.

Pobee, John S. "Religion and Politics in Ghana, 1972–1978: Some Case Studies from the Rule of General I. K. Acheampong." *Journal of Religion in Africa* 17, no. 1 (February 1987): 44–62. https://doi.org/10.2307/1581075.

Sanneh, Lamin. *Whose Religion Is Christianity?: The Gospel beyond the West*. Grand Rapids, MI: Eerdmans, 2003.

Slaymaker, William. "Ecoing the Other(s): The Call of Global Green and Black African Responses." *PMLA* 116 (2001): 129–144.

Thornber, Karen Laura. *Ecoambiguity: Environmental Crises and East Asian Literatures.* Ann Arbor: University of Michigan Press, 2012.

VanZanten, Susan. "World Christianity." In *The Cambridge Companion to Literature and Religion*, edited by Susan Felch, 262–276. Cambridge: Cambridge University Press, 2002.

Vital, Anthony. "Toward an African Ecocriticism: Postcolonialism, Ecology and *Life & Times of Michael K*." *Research in African Literatures* 39, no. 1 (Spring 2008): 87–106.

——— and Hans-Georg Erney (eds). "Postcolonial Studies and Ecocriticism." *Journal of Commonwealth and Postcolonial Studies* 13, no. 2 and 14, no. 1 (2006–7).

Walls, Andrew F. "Old Athens and New Jerusalem: Some Signposts for Christian Scholarship." *International Bulletin of Missionary Research* 21, no. 4 (October 1997): 146–153.

White, Lynn Jr. "The Historical Roots of Our Ecologic Crisis." *Science* 155 (1967): 1203–1207. https://doi.org/10.1126/science.155.3767.1203.

Yankah, Kwesi. "To Praise or Not to Praise the King: The Akan 'Apae' in the Context of Referential Poetry." *Research in African Literature* 14, no. 3 (1983): 381–400.

4 Waves and Refugees

Water Metaphors and Epistemological Humility in Thi Bui's *The Best We Could Do*

Erin Goheen Glanville

Introduction

This chapter takes as its conversation partner Deborah C. Bowen's article "Seeing Beyond the Scenery: Exploring the World Through Metaphor" (Bowen, 2013), which argues that metaphor is a marker of epistemological humility in poetry.[1] Widening that argument beyond the instance of poetry, this chapter asks what her conceptualization of metaphor helps us to understand about stale and fresh metaphors for human migration. Metaphor is commonly described as the overlapping of conceptual domains, which allows the communicator to draw from a source domain in order to understand a target domain better (see, for example, Lakoff, *Metaphors*, 2003). The imagery of water, the sea, and waves are prominent metaphors in academic and popular conversations seeking to understand forced migration. This makes sense in terms of lived experiences as well: many cultural texts have recognized the centrality of water in forced migration realities, and water is often the literal passageway for migration. Significantly, though, in most popular media, the conceptual domain this metaphor draws on is that of natural disaster or rising pressure.

In the following chapter, I consider the overuse of the conceptual domain of water disaster metaphors to frame the phenomenon of global migration. The effects of this overuse include a dead metaphor, or a fusion—one might say a rusted hinge—that no longer recognizes the "is not" of the metaphor. Following Bowen's thought, this metaphor thus suffers not only from a lack of creativity but from a lack of humility. And the lack of humility in this particular metaphor has serious implications: it dehumanizes and strips the *imago Dei* from image bearers around the globe. I offer Thi Bui's graphic memoir *The Best We Could Do* (Bui, 2017) as an example of how refugee authors can reclaim a restorative connection between images of water and human migration and to show how Bui finds fresh metaphorical frames through a return to epistemological humility. Humility is evidenced, in part, by her decision to rework metaphors from within the limits of her own family story—taking an emic approach. Emic narratives are stories told by those who speak from within an experience. This contrasts with etic perspectives, which attempt to produce an outside, objective perspective through intercultural comparison or cross-cultural observation. Bui's evocative metaphors for refugee lifeworlds

DOI: 10.4324/9781003366744-4

illustrate the creativity that can emerge when we choose to work explicitly within the limits of our own experiences.

I begin by revisiting Deborah C. Bowen's article on metaphor to establish an approach to metaphor that links to epistemological humility. I then provide a brief literature review of scholarship on the use of water metaphors in refugee discourse and popular communication about forced migration. Having established this as a dead metaphor, or a rusted hinge, I turn to Bui's memoir to understand how and what Bui reclaims from the domain of natural disasters to narrate her own family history. I conclude with a close reading of three striking images in her book, each of which makes a restorative "connection between the suffering human body and the body of the earth" (Bowen, "Seeing," 2013, 71). Taken together, these images provide readers with evocative visual metaphors for forced migration that demonstrate both the powerful potential of metaphor in story and the importance of humility as we reach for meaningful patterns with and within the limits of our visual and textual language.

Metaphor as Epistemological Humility

Bowen's article draws on authors as diverse as Suzanne Van Zantingh Gallagher and Roger Lundin, David Leary, Paul Ricoeur, Stephen Wright, Elaine Botha, and I.A. Richards, to establish metaphor as "an epistemological mode of inquiry" (Bowen, 2013, 63). With Psalm 19 as her first example, Bowen conceptualizes poetry as an "instance of metaphor as an epistemological mode" (65). Metaphors, she offers, are a hermeneutic for reading creation; they "give us the opportunity both to recognize our situation and to see—or hear—beyond it" (66). She proposes

> that imaginative literature, whether by Christians or not, may offer a way to recognize the voice of a non-human world that speaks without human language, through the power to place one thing in interaction with another that is both like it and unlike it.
>
> (66)

Her analysis is indebted to Jan Zwicky, who understands metaphor as "one way of showing how patterns of meaning in the world intersect and echo one another" (Zwicky, *Wisdom*, 6, qtd in Bowen, 66). For both Zwicky and Bowen, recognizing patterns is about convergence and divergence. In other words, metaphor points to two things at once: to what meaning finds echoes in other parts of creation and how those same patterns do not fit together (Bowen, 2013, 7). Thinking alongside Zwicky, Bowen spools out the significance of metaphor's *is* and *is not*, as its power to enact epistemological humility for the person who wishes to listen for that which exists beyond the limitations of language. The poetry of Don McKay and John Terpstra provides Bowen with concrete examples of this kind of humility as epistemology.

Multiple metaphors for metaphor are scattered throughout Bowen's article, but the image of a hinge is the one I wish to draw on for this chapter. Her reading of

Terpstra and McKay concludes with a subsection entitled "spiritual geography," which she begins in this way: "in their metaphors, these poets are enacting, rather than naming, a hinge between . . . the clay of the human and the clay of the earth, language and the 'beyond' of language'" (Bowen, 2013, 75). The image of a hinge stays with me as I read Bowen's summary of Adam Dickinson's idea, that "to describe the material world in metaphor" is to recognize that elements in the material world are most themselves "in their connectedness" (76). If metaphor is the overlap of two conceptual domains in order to see how patterns echo each other, the hinge exists to open small windows within that overlap. These windows provide us with new frames for seeing how the target domain *is not* an echo of the source domain.

According to Bowen, the hinge of metaphor is its ability to open up and reveal that which cannot be contained by human language, a spiritual reality that pulses in the gap between what something is and what it is not. The power of a hinge to rotate and reveal might refer also to the action of art as iconic vis à vis Madeleine L'Engle, who describes art as "an open window through which we can be given a new glimpse of the love of God" (L'Engle, "Icons," 59, qtd in Bowen, 64). Bowen concludes her article by naming this spiritual reality: what metaphors reveal to us is "something otherwise unnameable . . . something that we *might call* the glory of God in the material world" (78, emphasis added). Bowen's concluding remarks demonstrate her awareness of the tendency of conclusions to lack epistemological humility and to close down meaning. She uses a modalized verb (might) as a grammatical embodiment of both possibility and limitation, recognizing the "glory" of God also as a metaphor for that which exists beyond language.

The specific kinds of metaphors Bowen is interested in exploring are ones that help humans to hear the nonhuman creation speak. Bowen describes the anthropomorphizing of the earth in Terpstra's poetry as "an act of love." She echoes his own description of his work, calling it "a material kind of lectio divina, to make a connection between the suffering human body and the body of the earth" (Bowen, 2013, 72). The power of Bowen's own metaphor of poetic images as hinges is the way it brings together the conceptual domain of sensing through a static image with the conceptual domain of a moveable mechanical joint that turns on a fulcrum to open up a new line of sight. If, as Bowen suggests, the hinge of metaphor allows for the reader of poetry to see the sacred that is hidden within creation around, then it stands to reason that when the hinge is no longer moveable, that which could have been seen is no longer opened up through the use of that metaphor. To extend and perhaps mix the metaphors, I might describe the idea of a "dead metaphor" as a rusted hinge—a hinge that no longer functions as it was meant to, that has remained stationary so long in the elements that it has rusted shut.

In her article on the metaphor of "spiritual warfare" E. Janet Warren describes a metaphor that has "died" as potentially dangerous. She writes:

> The danger of having only one metaphor is that it can be overused; the original referent is forgotten, the metaphorical tension is lost, and the terms become part of ordinary language. This is known as a dead or conventional

> metaphor, common examples being the 'legs' of a table, a 'field' of research, or a 'blind' review.
>
> (Warren, "Spiritual Warfare," 2012, 286)

In other words, when we forget the usefulness or the purpose of overlaying two conceptual domains (that is, to open up new insights through a tension between what is and what is not similar), we risk fusing the lesser-known referent, for instance, "refugee," with the familiar metaphor, for instance, "rising water pressure" or "natural disaster." The potential for the hinge to swing open a new window for deeper sensing is lost; the hinge rusts shut. To unpack the implications of this, allow me to first establish and detail the rusted hinge we are exploring in this chapter.

Water Metaphors as a Metaphorical Frame for Refugee Lifeworlds

"What Is": The Use of Water to Describe Refugee Lives

In 2017, the well-known Chinese artist Ai Weiwei released his documentary *Human Flow*, which aimed to offer a sympathetic narrative of "the global refugee crisis" and which made enough of a splash to be shortlisted for the Oscars. Shots of people waiting on beaches, crying in boats on choppy water, and struggling to pull each other upstream reference a particular kind of relationship between water and many (not all) people seeking refuge—a relationship in which water is a place of both leaving and entering, access and threat, necessity and choice. Using drones to produce a large-scale aesthetic, *Human Flow* visually fuses this actual relationship to the common metaphorical framing of refugee people *as water*. The camera flies quickly over or beside miles of winding roads filled with people on the move, providing a birds-eye view of tens of thousands of people appearing two-dimensionally against the earth. Over and over, the film visualizes people as flows over the face of the earth—sometimes contained, sometimes bursting out of their containers. On *IMDb*, the film's storyline is littered with catastrophic words like "massive," "staggering," "urgent," "teeming," "perilous," "desperate," "crucial" (www.imdb.com/title/tt6573444/).

In fact, "refugees as water" may be the most persistently pervasive metaphorical frame used to describe refugee lifeworlds. Researchers have demonstrated its pervasiveness to be true—especially in news media—in numerous cultural, linguistic, and national contexts and, through historical analyses, across time (Charteris-Black, "Britain," 2006; Porto, "Water Metaphors," 2022; Taylor, "Metaphors," 2021). The frame encompasses metaphors that blend the forced migrant with liquids, such as floods, waves, streams, swells, and spills (Abid et al., "A flood," 2017; Porto, "Water Metaphors," 2022; Taylor, "Metaphors," 2021). Researchers largely agree that this metaphor negatively represents refugees as a threatening mass that must be contained (Abid et al., "A flood," 2017; Catalano, "When Children Are Water," 2017; Charteris-Black, "Britain," 2006; KhosraviNik, "The representation of refugees," 2009; Nguyen and McCallum, "Drowning," 2016; Porto, "Water Metaphors," 2022; Salahshour, "Liquid metaphors," 2016; Sinclair,

"*Agua*-Biographies," 2020). It often invokes notions of natural disaster, representing forced migrants as an overwhelmingly dangerous force (Abid et al., "A flood," 2017; Catalano, "When Children Are Water," 2017; Charteris-Black, "Britain," 2006). For example, Abid et al. discover that water metaphors are used by both host and non-host countries in news media about Syrian refugees and that these references reveal a primary concern: excessive volume ("A flood," 2017, 123, 127). In other words, the metaphor is used to frame forced migration as a large problem to be solved. Common metaphors centre on ideas of "flooding" and "hailing," presenting an influx of refugees as a burdensome disaster (Abid et al., "A flood," 2017, 138). Catalano ("When Children Are Water," 2017) concurs, arguing that, in Nebraska, the pervasive metaphor of "immigrant children are dangerous water" frames them as a relentless force that imposes pressure on the nation (6–7). They argue that blending characteristics of dangerous water with persons seeking refuge dehumanizes, homogenizes, and differentiates them as unwelcome outsiders (Catalano, "When Children Are Water," 2017, 5). In many news articles, the language of "swamping" is used to describe the effect of forced migration on health or social services. Focusing on the less easily quantified issue of national culture, Charteris-Black ("Britain," 2006) shows how natural disaster metaphors are used to express fear about excessive rapid change and the way refugees are framed as representative of that change (572).

Even when non-disaster water metaphors are used, some researchers affirm that they produce a largely negative frame for forced migrants (Porto, "Water Metaphors," 2022; Salahshour, "Liquid metaphors," 2016; Sinclair, "*Agua*-Biographies," 2020). Sinclair draws on Derrida to argue that water, in being intangible and diffuse, renders refugees unintelligible, lacking a solid place, time, or clear definition, lacking access to a particular nationality or autobiography ("*Agua*-Biographies," 2020, 345, 360). Porto ("Water Metaphors," 2022) examines the underlying "image schemas" behind words such as "flow" to argue that they convey negativity. "Flow" contains repeated image schemas of "force," and when paired with image schemas of host countries as a "container," suggest that refugees present a threat (259–261). In fact, many authors contend that the metaphor of "refugees are water" is invariably tied to "country is a container," and these must be analysed together to discern full meaning (Charteris-Black, "Britain," 2006; KhosraviNik, "The representation of refugees," 2009; Nguyen and McCallum, "Drowning," 2016). Some posit that simply conceptualizing refugees as nonhuman implies evaluation, rendering the refugee an object without political agency (Catalano, "When Children Are Water," 2017; Charteris-Black, "Britain," 2006; Porto, "Water Metaphors," 2022; Taylor, "Metaphors," 2021).

That said, there is not complete consensus on the significance of the water metaphor as a frame for refugee lifeworlds. Some authors, though in the minority, have voiced exceptions to an overarching negative significance. KhosraviNik ("The representation of refugees," 2009), Nguyen and McCallum ("Drowning," 2016), and Salahshour ("Liquid metaphors," 2016) have emphasized the importance of contextualizing the metaphor within the social, political, geographical, and historical circumstances in which they emerge. For example, water metaphors in

Britain describing the volume of refugees during the Balkan Crisis served to generate sympathy, if understood in the "macro-structure" of the time which emphasized humanity and humanitarian aid (KhosraviNik, "The representation of refugees," 2009, 486). Contrastingly, Sinclair posits that the water metaphor is generally negative, as Western metaphysics privileges "solidity, presence, fixity, and substance" ("*Agua*-Biographies," 2020, 345). However, they argue that a decolonial framing would embrace elements of liquid that challenge the West's captivation with solidly defined place and citizenship (Sinclair, "*Agua*-Biographies," 2020, 360–361).

"What Is Not": The Impact of a Rusted Hinge, or Dead Metaphor

While researchers are aware of the metaphorical status of water in relation to refugees, its near-ubiquitous association suggests that in everyday communication about refugees as "waves" we are no longer aware of the metaphor. I have experienced this in community workshops where I am critiquing the metaphor and then accidentally use it for want of an alternative. The phrase "waves of refugees" communicates mastery through language—the writer's ability to categorize a vast array of human experiences as a single phenomenon. And it appears to function without any openings for what "is not" similar because the metaphor does not operate at the level of metacognition. The rusted hinge of this metaphor closes off the potential for communicating with humility about what is not known in forced migration.

We might wonder what danger there is in the rusted hinge: does a dead metaphor merely signify a lack of fresh insight or creativity? Granted, I have never heard an outcry about the ethics of a dead metaphor like "table legs." Yet, the metaphor of rising water pressure or the natural disasters of water to describe refugees—perhaps because it describes human beings—clearly does more than shut down the opportunity for fresh insight. An accumulation of metaphors that work in a similar direction builds a frame for topics or issues. Lakoff writes, ". . . frames are in the synapses of our brains—physically present in the form of neural circuitry" (*Don't Think of an Elephant*, 2004, 73). Once we have a particular frame in the synapses of our brains, it can be difficult to assimilate any facts that contradict that frame. People will discard a fact if it does not fit their frame for that conceptual domain. Metaphors accumulate to build frames; frames determine how we assess the truthfulness of facts; and so, the use of metaphors in public communication about refugees builds a frame that filters facts.

In other words, the metaphor of refugees as natural disasters operates as linguistic and neural infrastructure for an active debate about how countries such as Australia, Britain, Canada, or the United States should respond to migration. Migration as impending crisis frames a particular way of thinking about not only what global migration is but what it calls for from me as a citizen in a wealthy and relatively secure country. The use of this dead metaphor is constitutive of a limit, then, on the horizons of a person's imagination for what they might do in response. And at least one standard response to rising water pressure is protection against it, planning and acting to mitigate its risk, and gathering information about the risk potential in order to manage it. If we insert "a person seeking refuge" in place of "it," we begin to see the implications. The effects of rising water pressure as a metaphor for refugee movement are both conceptual and empirical. The merging of climate crises and refugee crises at this point in history produces an imagined future in which poor people are displaced from places where the earth has been devastated by rising water. And taken together, they constitute a double-barrelled threat to the security of rich citizens in powerful nation-states.

So where does one go from here? What is interesting in the brief review of literature offered earlier is the way that scholarship focuses almost exclusively on the conceptual significance of the water metaphor when applied to the refugee, and not on the significance of water for those seeking refuge. In other words, while research has thoroughly explored the socio-linguistic construction and consequences of the pervasive water metaphor, what remains unexplored is an assessment of how salient this metaphor and/or its referent is for forced migrants themselves. We are missing an emic narration of water for refugee life.

In the remaining sections, I turn to Thi Bui's graphic memoir *The Best We Could Do* to understand how Bui reclaims water as a non-metaphorical, distinct part of creation in her own telling of her family's experience of displacement. Bui brings the physical reality of water to life through her visualizations of actual bodies of water and by choosing watercolour as her medium. In this way, her memoir flips the metaphor so that refugee lifeworlds become the source domain that can reframe the significance of water as a target domain. Further, Bui offers a genuinely fresh metaphor for refugee lifeworlds in the form of a family tree, uncommonly illustrated to produce a new hinge for seeing into the sacred space of displacement within family relations.

Reviving Metaphor Through Epistemological Humility: Thi Bui's *The Best We Could Do*

Thi Bui took 12 years to complete her first graphic memoir, *The Best We Could Do* (*TBWCD*) (2017). This in itself speaks to the consideration required to visualize and write about human displacement, but her interviews and preface also speak to the care she took in representing communal and familial history. *TBWCD* is an international bestseller and has been praised especially for offering a woman-centred narration of forced migration (Drews, "Something I Have Created," 2022; Oh, "Birthing," 2020) and for challenging the stereotypes of Vietnamese people produced in movies about the Vietnam War (Gusain and Smita, "Trauma," 2021; McWilliams, "Precarious Memories," 2019). *TBWCD* tells the story of Bui's journey to understand her family's experiences as Vietnamese refugees who were resettled in the United States in the 1970s. Water features throughout.

Resisting Popular Uses of the Water Metaphor

Unsurprisingly, extratextual evidence provides our first clue to understanding Bui's use of water metaphors. On the cover, Bui's family is painted red, backs to the reader, standing out against the blue watercolour of the ocean that separates them from their future and their past homes. They are visualized as, most decidedly, *not the waves*. Rather, they stand on the shore facing those waves, which become an ambiguous passageway and natural threat to *their* safety in the search for home. The clear distinction between human migration and rising water pressure is established in part by the ability of a visual medium to present images as simultaneously literal and representative of something else. If we return to our conversation about standard responses to rising water pressure, we now see a refugee family considering how to find protection across the water, planning and acting to mitigate the risk of displacement, and gathering information about the risk potential in order to survive. Starting with the book cover, Bui's medium of graphic memoir breaks open this rusted textual metaphor by flooding it with the distinct reality of water and with a multitude of uncommon connections between refugee lifeworlds and water.

One of the critiques of water as a metaphor for global refugee movement is its dehumanizing scope: "waves of migration" represents massive movements of large, unnamed masses who are then enumerated as "64 million refugees worldwide." The "humanizing" alternative in humanitarian communication or news media is the human-interest story of an individual refugee, focusing in on their personal victimhood and tale of rescue. This approach has been critiqued in turn for its depoliticization of refugeehood. In contrast to either of these more standard rhetorical approaches (dehumanizing statistics or depoliticizing personalized realism), Bui writes in her preface that the challenge of the book was "how to present history in a way that is human and relatable and not oversimplified" (Bui, *The Best We Could Do*, 2017, n.pg.). Bui accomplishes this by both widening and narrowing the frame for refugee lifeworlds in unexpected directions.

First, she widens the frame. Before the story begins, readers are offered a two-page splash that outlines the timeline of her country's colonial history. We follow the events as they rise and fall in the shape of waves across the pages, and thus are invited to frame Bui's family memoir within the grand scope of 2,000+ years of political history. If we are to understand her family's story, we must read it as a result of geopolitical violence across millennia; successive waves of imperial and colonial power are the waves that have threatened security. The security that has been threatened is ultimately that of their family.

TBWCD is no depoliticized human-interest story. But the timeline is immediately followed by the first page, titled "Labor." This is where she narrows the frame. We see a single panel, contained by a heavy border, where readers are given a jarringly intimate perspective: we look down at Bui's pregnant belly, as she prepares to give birth. Evoking the idea of water immediately in the breaking of water out of her body, the text holds this literal water in tension with another water metaphor. We read Bui's thoughts: "The pain comes in twenty-foot waves and Ma has disappeared" (Bui, 2017, 1). Here water is a metaphor for pain that washes over and through the female refugee body and links to the loss of maternal connection between displaced generations. By the end of this first chapter, the wave of pain has shifted into "a wave of empathy" that "washes over" Bui as she imagines her mother giving birth in much more difficult circumstances (22). In this conceptualization of refugee lifeworlds, the water of the earth is also the water that breaks out of women's bodies before birth, and the metaphorical wave for that moment is of pain and later empathy.

The big picture of geopolitics retains political power as the widened frame for the story, and the immediate narrowing to the deeply personal moment of giving birth tells the reader that migration is about human beings and their unique experiences. In both the political and the personal, water is evoked as simultaneously metaphorical and literal, thus beginning a process of decoupling both water metaphors from natural disasters and water more generally from people seeking movement. But, as we see in the following section, to refresh this metaphor, Bui needs to dig down into the details of her own family story.

Reclaiming Water Metaphors via Refugee Lifeworlds

Water runs through the entire memoir, as references to real bodies of water and as a metaphor for what stands between Bui and a sense of home. Bui's lack of connection to a homeland is established in one especially provocative image that layers geographical displacement onto the suffering human body. As we turn to page 36, Bui has just decided to begin recording her family history "thinking that if I bridged the gap between the past and the present . . . I could fill the void between my parents and me" (36). In the top lefthand corner of the page, she has drawn herself drawing, sitting against the backdrop of a breaking wave. We follow the watercolour as it seeps out of the frame to form the shape of Vietnam's geographical border. Standing in the foreground, looking at it, is Bui as a child, her back to the reader, hair whipping around in the wind, vulnerably naked. The reader can see through to the other side of her; a hole has been cut out of her core, and the hole is in the shape of Vietnam. Her hands reach tentatively into the gap where her heart and lungs should be, and the southern tip of the country draws our eyes down along the length of her missing spine. In the text boxes, we read: "And that if I could see Viet Nam as a real place, and not a symbol of something lost . . . I would see my parents as real people . . . and learn to love them better" (36). Bui reaches into an internal abyss searching for the layered grounding of place and family. Symbols of home have been inadequate for solidifying her sense of identity; she needs to sense that homeland herself.

This search for a homeland is necessitated by the gaps in her own memories (what has been termed "post-memory" by Holocaust scholar Marianne Hirsch), the intergenerational trauma that shapes her own emotional and psychological landscape, and the silences of her parents. For Bui and other 1.5-generation refugees, to find a home is to gather second-hand memories from the older generation, to sift through the etic stories of historical archives and pop culture references, and to return to unfamiliar physical landscapes and integrate contemporary sounds, smells, and tastes into an intangible past. Bui does all this. She traces their "journey in reverse . . . over the ocean through the war seeking an origin story that will set everything right" (40–41). But the origin story she seeks does not exist.

Bui's resolution comes, not from finding an origin story, but in a recognition of the shared vulnerability of land and humanity. In her final chapter, she writes:

> At least I no longer feel the need to reclaim a HOMELAND. I understand enough of Việt Nam's history now to know that the ground beneath my

> parents' feet had always been shifting . . . so that by the time I was born, Việt Nam was not my country at all. I was only a small part of it.
>
> (326)

Being a small part of something liquid is, ironically, a relief for Bui. But this relief does not take away her worry: "What has worried me since having my own child was whether I would pass along some gene for sorrow or unintentionally inflict damage I could never undo" (326–7). Here Bui is referring to discussions about intergenerational trauma (what she calls the "refugee reflex" for survival) and its potential to be passed down in the DNA of refugee families. The determinism of this idea disturbs her until she finds a potential answer in a memory from childhood.

When she was young, Bui saved up money to give to her mother for Mother's Day. "But when I counted the coins, I found that my math had been all wrong—and that I was quite a bit short. Instead of a triumphant gift, Ma got a wet shoulder from my tears" (Bui, *The Best We Could Do*, 2017, 320). Her mother comforts her with grace and a gentle reminder that failure is okay: "It's enough that you thought of me," she says. Bui reflects on this memory in relation to her disappointment with her parents. "What if all my mother had remembered was that I came up short? . . . I wasn't ready to lose my mother when I was thirteen but now, at forty, I know that our time on earth is finite" (322). To use Bowen's language, Bui finds her answer in a kind of epistemological humility as she acknowledges what she does not know—in both the past and the present—and what she cannot reclaim—in herself or in her parents. In fact, not knowing her mother's traumas, though they have impacted Bui deeply, and not knowing how the grief of displacement will be passed down to her son actually create a space for contentment and even gratitude in the final chapter.

Making a New Metaphor Out of Refugee Lifeworlds

Bui's discovery of a more fluid sense of identity does not erase her search for roots. Scholars in cultural refugee studies have not explored the potential for refugee lifeworlds to be framed through the metaphors of trees, roots, or plants. This may be in part because flora has not been a dominant frame to conceptualize the experiences of forced migrants in public discourse, though there have been other contexts in which it has shown up (Abid et al., "A flood," 2017; Catalano, "When Children Are Water," 2017; Charteris-Black, "Britain," 2006; KhosraviNik, "The representation of refugees," 2009; Nguyen and McCallum, "Drowning," 2016; Salahshour, "Liquid Metaphors," 2016; Taylor, "Metaphors," 2021). For example, the "Tree of Life" metaphor has been applied in mental health interventions to support refugee youth in describing their life experiences (Hughes, "Finding a Voice," 2014; Jacobs, "Collective Narrative," 2018). Within this metaphor, "ground" has symbolized the refugee's current environment; "roots," their origin; "trunk," their abilities; "branches," their hopes; "leaves," their loved ones; and "fruits," gifts they have received (Jacobs, "Collective Narrative," 2018, 285–288). This therapeutic framework finds some company in social science's language of systems change, where the focus is on discovering the "roots" of forced migration.

Roots, specifically, have been explored as a metaphorical frame for the experience of forced displacement in literature, such as in Alex Haley's 1976 novel *Roots* (Wampole, *Rootedness*, 2016, 3). While not pervasive in popular depictions of forced migration, notions of rootedness have been a long-standing frame for human engagement with the world (3). Wampole explores the evolution of the root metaphor in her book *Rootedness* (2016), revealing its connection to exile. Critics of this metaphor argue that such a frame may fall into the Western metaphysical trap of privileging solidity, failing to embrace the fluid identities of refugees that challenge colonial understandings of place and citizenship (Sinclair, "*Agua*-Biographies," 2020, 345). While this critique bears consideration, Wampole argues that "when home is destroyed, left behind, or conquered, the urgency to feel rooted increases exponentially," suggesting a rich potential in the root metaphor for framing refugee lifeworlds (Wampole, *Rootedness*, 2016, 23).

Wampole's insight tracks with my own scholarship: I wrestled with the tension between rootedness and fluidity when trying to read refugee fiction through the lens of diaspora theory. "For poststructuralist-inflected postcolonial and diaspora theory, migrancy provides a model for resisting essentialism and cultural nationalism" (Glanville, "Re-routing diaspora theory," 2009, 129). Yet in Canadian refugee fiction,

> characters are less concerned with the essentializing nature of rootedness, less celebrative of routes, and more concerned with the obstacles that keep them from re-rooting themselves. Or, to use another metaphor, it is not the attachments to home that concern refugee characters in Canadian literature, but rather, the dramatic consequences of various 'reroutings', among which the loss of home and their desire to be at home.
>
> (Glanville, "Re-routing diaspora theory," 2009, 129)

In other words, any theorization of forced migration that reacts against the idea that rootedness is normative for human life must also account for memoirs of refugee lifeworlds that articulate a longing for old or new roots in dislocation.

Bui certainly wrestles with dislocation and her desire for home and rootedness. Immediately following her childhood memory of failure, she offers roots as a fresh metaphor for refugee lifeworlds, but with a twist. She visualizes her own family as a root system; the two-page spread displays a tree's root system overlaid with drawings of family members, like a reverse family tree (Bui, *The Best We Could Do*, 2017, 324–325). And the root systems morph across the page into veins, with round objects flowing through them like blood cells. As before, the suffering human body (this time communal) is layered onto the suffering body of creation. Where before the overlap presents as a missing core, this time it produces a shared sense of finitude and grace-filled limits. The fixed quality of tree root systems joins the liquid quality of the human venous system. The metaphor that emerges from refugee lifeworlds is a hybrid of liminality and groundedness. Together the root and venous systems speak of limits—both the limits of metaphors and the limits of creation. She admits to the reader that her search for roots came out of a need to be something more than she was:

> I used to imagine that history had infused my parents' lives with the dust of a cataclysmic explosion . . . that it had seeped through their skin and become part of their blood. That being my father's child, I too, was a product of war . . . and being my mother's child, could never measure up to her.
>
> (324–325)

In her words, I hear that even the experience of suffering can be framed as exceptional, as more-than-human, as beyond the limits of creation.

Like Bowen, Bui hedges the conclusions she has come to. She acknowledges, "But maybe being their child simply means that I will always feel the weight of their past. Nothing that happened makes me special. But my life is a gift that is too great—a debt I can never repay" (Bui, *The Best We Could Do*, 2017, 325). Again, Bui is speaking into conversations about refugee representation regarding the weight of gratitude that refugees are asked to bear amid loss (see, for instance, Dena Nayeri's "The Ungrateful Refugee"). Instead of resenting the gratitude that is expected of her, Bui reframes gratitude as something she owes to her parents, not to the state that offered protection. Along with this reframing, she gives herself a new expectation: namely, to fail at repaying the debt. Thus, partial knowledge of one's small ordinariness and even failure becomes freedom.

Conclusion

Bui wishes to release her son also from the weight of survival and the indebtedness it produces, and so her final pages return to him in two water scenarios (Bui, *The Best We Could Do*, 2017, 328–9). In the first, he is splashing in the ocean playfully, reminiscent of her own playful splashing in a refugee camp (260, 270). In the second, he is swimming away underwater with the bubbles of his breath floating above him, reminiscent of her childhood daydreams in the shadow of her father's trauma (89). Viewing these two scenarios from within a framed inset, an adult Bui reflects, "When I look at my son . . . I don't see war and loss or even Travis and me. I see a new life, bound with mine quite by coincidence, and I think maybe he can be free" (328). These two final images return us to the work Bui is doing with the metaphor of water for refugee lifeworlds. Once the target and source domains are flipped, refugees become the source of knowledge for understanding the significance of water. Water is the stuff of death and PTSD nightmares; it makes possible birth and play and sustains life; it is both a beginning and an ending, a passageway and a wall; it points to possibilities for freedom and a lack of rootedness. Providing us with a new emic metaphor, Bui brings together the root systems of a tree with the play of a child in the ocean to open up the potential for healing in the overlap of human displacement and the earth—trees, water, and land.

In these fresh metaphors, the nonhuman creation speaks, and this reader recognizes "the voice of a non-human world that speaks without human language, through the power to place one thing in interaction with another that is both like it and unlike it" (Bowen, "Seeing," 2013, 66). Bui's son splashes and breathes, creating bubbles in the ocean, and these pockets of air serve as a reminder that the water metaphor need not be all-encompassing. Merging the bodies of trees and the bodies of people, Bui's visual metaphor reveals a supportive root system that passes on not only intergenerational trauma but also intergenerational survival and love. Her conclusion points to more than a reclamation of agency. It also points to the freedom that comes from knowing we have done "the best we could do" in our relationships to one another and to the earth, the gratitude, even, that we might experience as finite people navigating loss on a suffering globe. Bowen *might* call this the beauty of epistemological humility: as we immerse in these visual metaphors where human and nonhuman creation overlap, we begin to sense the sacred that pulses beyond what is.

Note

1 Thanks to Anjali Mishra for background research assistance on this chapter.

References

Abid, Raith, Shakila Manan and Zuhair Rahman. "A Flood of Syrians Has Slowed to a Trickle: The Use of Metaphors in the Representation of Syrian Refugees in the Online Media News Reports of Host and Non-Host Countries." *Discourse and Communication* 11, no. 1 (2017): 121–140. https://doi.org/10.1177/1750481317691857.

Ai, Weiwei. *Human Flow*, 2017. Documentary film. www.imdb.com/title/tt6573444/

Bowen, Deborah. "'Seeing Beyond the Scenery': Exploring the World Through Metaphor." *Canadian Theological Review* 2, no. 1 (2013): 59–78.

Bui, Thi. *The Best We Could Do*. New York: Abrams Comicarts, 2017.

Catalano, Theresa. "When Children Are Water: Representation of Central American Migrant Children in Public Discourse and Implications for Educators." *Journal of Latinos and Education* 16, no. 2 (2017): 124–142. https://doi.org/10.1080/15348431.2016.1205988.

Charteris-Black, Jonathan. "Britain as a Container: Immigration Metaphors in the 2005 Election Campaign." *Discourse & Society* 17, no. 5 (2006): 563–581. https://doi.org/10.1177/0957926506066345.

Drews, Marie. "'Something I Have Created': Breastfeeding and Motherhood Trauma in Thi Bui's *The Best We Could Do*." *Literature and Medicine* 40, no. 1 (2022): 121–146. https://doi.org/10.1353/lm.2022.0012.

Glanville, Erin Goheen. "Re-routing Diaspora Theory with Canadian Refugee Fiction." In *Rerouting the Postcolonial: New Directions for the New Millennium*, edited by Sarah Lawson Welsh, Cris Sandru and Janet Wilson, 128–138. London: Routledge, 2009.

Gusain, Abhilasha and Smita Jha. "Trauma, Memory, History and Its Counter Narration in Thi Bui's Graphic Memoir *The Best We Could Do*." *Journal of Graphic Novels & Comics* 12, no. 5 (2021): 873–884. https://doi.org/10.1080/21504857.2020.1806890.

Hughes, Gillian. "Finding a Voice Through 'The Tree of Life': A Strength-Based Approach to Mental Health for Refugee Children and Families in Schools." *Clinical Child Psychology and Psychiatry* 19, no. 1 (2014): 139–153. https://doi.org/10.1177/1359104513476719.

Jacobs, Suzan F.M. "Collective Narrative Practice with Unaccompanied Refugee Minors: 'The Tree of Life' as a Response to Hardship." *Clinical Child Psychology and Psychiatry* 23, no. 2 (2018): 279–293. https://doi.org/10.1177/1359104517744246.

KhosraviNik, Majid. "The Representation of Refugees, Asylum Seekers and Immigrants in British Newspapers During the Balkan Conflict (1999) and the British General Election (2005)." *Discourse & Society* 20, no. 4 (2009): 477–498. https://doi.org/10.1177/0957926509104024.

Lakoff, George. *Don't Think of an Elephant! Know Your Values and Frame the Debate*. White River Junction, VT: Chelsea Green Publishing, 2004.

——— and Mark Johnson. *Metaphors We Live By*. 1980. Chicago: University of Chicago Press, 2003.

McWilliams, Sally. "Precarious Memories and Affective Relationships in Thi Bui's *The Best We Could Do*." *Journal of Asian American Studies* 22, no. 3 (2019): 315–348.

Nguyen, Li and Kerry McCallum. "Drowning in Our Own Home: A Metaphor-Led Discourse Analysis of Australian News Media Reporting on Maritime Asylum Seekers." *Communication Research and Practice* 2, no. 2 (2016): 159–176. https://doi.org/10.1080/22041451.2016.1188229.

Oh, Stella. "Birthing a Graphic Archive of Memory: Re-Viewing the Refugee Experience in Thi Bui's *The Best We Could Do*." *MELUS: Multi-Ethnic Literature of the U.S.* 45, no. 4 (2020): 72–90. Project MUSE.

Porto, M. Dolores. "Water Metaphors and Evaluation of Syrian Migration: The *Flow of Refugees* in the Spanish Press." *Metaphor and Symbol* 37, no. 3 (2022): 252–267. https://doi.org/10.1080/10926488.2021.1973871.

Salahshour, Neda. "Liquid Metaphors as Positive Evaluations: A Corpus-Assisted Discourse Analysis of the Representation of Migrants in a Daily New Zealand Newspaper." *Discourse, Context & Media* 13 (2016): 73–81. https://doi.org/10.1016/J.DCM.2016.07.002.

Sinclair, Rebekah. "*Agua*-Biographies: Derrida on Water, Ontopology, and Refugees." *The Journal of Speculative Philosophy* 34, no. 3 (2020): 353–366. https://doi.org/10.5325/jspecphil.34.3.0353.

Taylor, Charlotte. "Metaphors of Migration Over Time." *Discourse & Society* 32, no. 4 (2021): 463–481. https://doi.org/10.1177/0957926521992156.

Wampole, Christy. *Rootedness: The Ramifications of a Metaphor*. Chicago: University of Chicago Press, 2016. https://doi.org/10.7208/chicago/9780226317793.001.0001

Warren, E. Janet. "'Spiritual Warfare': A Dead Metaphor?" *Journal of Pentecostal Theology* 21, no. 2 (2012): 278–297. https://doi.org/10.1163/17455251-02102007.

5 Poems by John Terpstra (I)

Ceremony

Some ceremony is involved
in hauling the bicycles up from the basement,
angling their awkwardnesses
under ductwork, around the stairwell landing,
and through the needle's eye of the side door,
as one wheel spins free, ticking.

The sun was singing down, loud and glorious,
as we wove our two thin tire lines through town
and to the bay, then through and around surprising
numbers of walkers, strollers, roller-
bladers of varying degrees of proficiency, and other
cyclists, all of us twisting and braiding
our various momentums together
like a fat rope played out by the water's edge,

and the gravel swallowed our tires as we left the paveway
and headed toward the point, while volleyed
in and out of hearing on the wind we heard the beat
of some hydraulic jack driving posts into the earth,

or some jack's boom box, I said,

but came upon the backs of three white
t-shirts there instead, and the draped jet
hair and braids of two men and a woman
sitting facing the open blast
furnace of light on the water,
in the sky, and drumming
the sun down,
singing rhythmic undulations over
the chanted surface of the bay,

DOI: 10.4324/9781003366744-5

a song that went nowhere, musically,
and could have gone there forever—and the gravel
swallowed our tires again, as we quit
the landfill point to be home before dark,
and all I wanted to convey is the deep enjoyment
of our early summer evening ride, the unexpected
explosion of people at the new waterfront park
three generations after industry and E.coli

drove them all away, and the three first-peoples,
this late in the game, chanting and drumming
the sunset as an event
in which they were participant,
while we looked on.

* Previously published in *Disarmament* (Gaspereau Press, 2003) and reprinted here by kind permission of Gaspereau Press

Invasive Species

They are not native,
the Queen Ann's Lace, chicory
and birdsfoot trefoil
that add so much to the country drive,
as they wave back and forth
from the shoulder of the road.

Their success, I have read,
is measured in how well they adapt
to harsh conditions,
how they thrive
in adverse circumstances.

When my parents invaded
from overseas,
spilling like bilge, like zebra muscles,
some would say,
from the hold of a boat,
they lived in a driving shed
attached to a barn.
The farmer, and the farm work,
were brutal. My parents did better
than survive.

The nature-guide further states
that non-native trees
may add a green splash to the landscape,
but do not attract insects,
which are adapted to native trees,
and therefore do not interest
or attract birds.

I stand here, a non-native son,
though born on this soil,
and walk my walk,
banging through the woods
cowbell around my neck
head in a cloud of bugs,
birds nesting in my beard,
searching for a place that will call me home.

* Previously published in *Mischief* (Gaspereau Press, 2017) and reprinted here by kind permission of Gaspereau Press.

Acknowledgement Close to the Ground

We are gathered on the traditional territory
of people who live close to the ground,

who touch and are intimate with
this patch of earth, its waters;

who know and name the many more, and wild, animals
    with whom they share life and death here,
    as sisters and brothers;

who follow its creeks
and dwell in the beauty of this place
    in ways we only imagine,

bound as we are to the streets of the city,
    and to the paths of settlement.

6 "Outrage From Lifeless Things"

Theodicy and the Anthropogenic Effects of the Fall in *Paradise Lost*

Ben Faber

Arguably the most famous lines from John Milton's *Paradise Lost* come at the end of the opening invocation, where the narrator declares that the primary purpose of the poem is theodicy—the demonstration of the justness of God:

> That to the height of this great argument
> I may assert eternal providence,
> And justify the ways of God to men.
> (Milton, *Paradise Lost*, ed. Dobranski, 2022, 1.25–26)

The ambiguity of the prepositional phrase "to men" suggests that theodicy is a matter of both *justifying to men* the ways of God and justifying *the ways of God to men.* Milton appeals to the reader's reason and emotions, directly and indirectly, in making the argument that Adam and Eve are fully responsible for failing the one test of their obedience not to eat of the Tree of Knowledge of Good and Evil. God's foreknowledge of the outcome, the poem argues, does not constrain the free will of the protagonists. As God the Father declares in Book 3, "I made him just and right,/Sufficient to have stood, though free to fall" (3.98–99). One could argue that, although this may be true of Lucifer and of prelapsarian Adam, the inferior nature and subordinate status of Milton's Eve (see 4.295–308) may have rendered her less sufficient to withstand temptation than Lucifer and Adam. At the same time, however, Eve's culinary choices when preparing a meal for Adam and their angelic guest (5.331–349) demonstrate that she is capable of temperance and moderation with regard to physical appetite; moreover, she is portrayed as sharing with Adam in the divine image and imprint of reason (4.291–293). In short, *Paradise Lost* teaches the reader that those who are capable are culpable; God is exonerated from blame for the fall of Satan, Adam, and Eve, and he is justified in their respective punishments.

But is God also just in punishing *creation* for the disobedience of Adam and Eve? Immediately after God renders his judgment on Adam and Eve in Book 10, the angelic chorus responds with enthusiastic approbation of his actions:

> He ended, and the heav'nly audience loud
> Sung hallelujah, as the sound of seas,

DOI: 10.4324/9781003366744-6

> Through multitude that sung: "Just are thy ways,
> Righteous are thy decrees on all thy works;
> Who can extenuate thee? . . ."
>
> (10.641–645)

In glossing these lines, editors typically direct readers to Revelation 19:6 for the image of the sounding seas, and to Revelation 15:3, 16:7, and 19:2 for the vindication of God's decrees:

> And they sung the song of Moses the servant of God, and the song of the Lamb, saying, Great and marvelous are thy works, Lord God Almighty: just and true are thy ways, King of Saints. . . . And I heard another out of the Sanctuary say, Even so, Lord God almighty, true and righteous are thy judgments. . . . For true and righteous are his judgments. . . .
>
> (Geneva New Testament, 1611)

In a way that is typical of his reworking of biblical texts, Milton conflates these three passages from Revelation into one, making subtle but significant changes in the process. The phrases "just and true are thy ways" and "true and righteous are thy judgements" become in *Paradise Lost* 10.643–4 "Just are thy ways,/Righteous are thy decrees on all thy works." The omission of "true" (which connotes "real, ideal, genuine" according to Strong's Concordance, G228) can be explained by the fact that this attribute is less relevant to Milton's project of theodicy in *Paradise Lost*. But the addition of a preposition is a significant alteration that Milton makes to the biblical source: rather than "marvelous are thy works," Milton has the angels sing "Righteous are thy decrees *on* all thy works" (emphasis added). Similar to the ambiguity of the preposition "to" in the poem's opening invocation ("justify the ways of God to men"), the reworking of the angels' song invites the reader to consider not only *God's works* in general but especially his decrees *on all his works*. In Milton's wording, "works" indicates more than God's ongoing providential engagement in history (i.e. his working) to include the creation itself and his pronouncements or judgments "on" all that he has made.

That the poem invites the reader to consider theodicy in relation to the punishment of creation is clear from the detailed description of cosmological changes in the following 63 lines of Book 10. Beginning with the sun, moon, and stars, the narrator works his way down the chain of being to describe existential alterations in the natural world, including "Outrage of lifeless things" and the "fierce antipathy" among animals:

> The sun
> Had first his precept so to move, so shine,
> As might affect the Earth with cold and heat
> Scarce tolerable, and from the north to call
> Decrepit winter, from the south to bring
> Solstitial summer's heat. . . .
> . . . .

Some say he bid his angels turn askance
The poles of Earth twice ten degrees and more
From the sun's axle; they with labor pushed
Oblique the centric globe. . . .
. . . .

At that tasted fruit
The sun, as from Thyéstean banquet, turned
His course intended; else how had the world
Inhabited, though sinless, more than now,
Avoided pinching cold and scorching heat? . . .
. . . .

Thus began
Outrage from lifeless things; but Discord first
Daughter of Sin, among th'irrational
Death introduced through fierce antipathy:
Beast now with beast gan war, and fowl with fowl,
And fish with fish; to graze the herb all leaving,
Devoured each other; . . .
(10.651–656, 668–671, 687–691, 706–712)

Aware of these "growing miseries" all about him, Adam laments, "O miserable of happy! Is this the end/Of this new glorious world . . . ?" (10.720–721). Here the theological concerns of *Paradise Lost* merge with ecological concerns that merge with eschatological concerns. Adam's misery is spiritual and psychological, as his interpretation of the signs of climate change leads to his sense that the consequence of his actions spells the demise of this brave new world. Readers today are often struck by Milton's prescient suggestion of a multiverse in the earlier books of *Paradise Lost* (see, for instance, 3.565–6, 8.148–9) that anticipates the images of distant galaxies now visible with the James Webb Space Telescope.[1] On the other end of the spectrum of creation, the Archangel Raphael touchingly praises "The parsimonious emmet" (7.485) in his vividly detailed account of creation. These two instances of Milton's affinity with natural philosophy—from the cosmological to the entomological—appeal to the ecocritical reader, both for the poem's interest in science and for the reader's analysis of early modern attitudes towards the natural world.[2] But the account of the heavy change to creation in the lines from Book 10 quoted earlier has a particular resonance with undergraduate readers living in the "Scarce tolerable" (10.654) climate conditions of the twenty-first century. In particular, students who frame their environmental advocacy and activism in terms of Christian stewardship and creation care will read the post-Fall books of *Paradise Lost* with a view to its "great argument" (1.24), namely, theodicy. Readers living in the Anthropocene, familiar with the notion of anthropogenic climate change, will wonder about the justness of God's actions in effecting collateral catastrophic damage to a creation that is innocent of wrongdoing. The poem may have justified the ways of God to humanity, but does it succeed in justifying the ways of God to creation itself?[3]

I would like to consider this question from the perspective of an undergraduate reader of *Paradise Lost* by exploring, first, the attributes of the good reader that Milton implies in the poem; second, the nature of creation; third, the environmental cost of the Fall; and, fourth, the justness of anthropogenic change: can the existential changes to the natural world, consequent to the fall of Adam and Eve, be justified within the moral and ontological framework of the poem? In the conclusion, I show how the question of theodicy and climate change offers a new way into *Paradise Lost* in the undergraduate classroom.

The Miltonic Reader

In *Areopagitica: A speech of Mr. John Milton for the Liberty of Unlicenc'd Printing, to the Parlament of England* (1644), Milton makes the rationalist argument that reading heterodox texts is a needful, healthy exercise in discernment:

> . . . what wisdom can there be to choose, what continence to forbeare, without the knowledge of evil? He that can apprehend and consider vice with all her baits and seeming pleasures, and yet abstain, and yet distinguish, and yet prefer that which is truly better, he is the true warfaring Christian. . . . That which purifies us is triall, and triall is by what is contrary.[4]
>
> (Milton, *Areopagitica, Prose Works*, 1959, 514–515)

In *Surprised by Sin: The Reader in* Paradise Lost, Stanley Fish shows that Milton's epic seeks to do precisely this—purify the reader by trial as they progress through contrarieties in the work:

> The poem's method is to involve you in its plot by confronting you with interpretive crises . . . the true center of *Paradise Lost* is the reader's consciousness of the poem's personal relevance, and that the arc of the poem describes, in addition to the careers of the characters, the education of its readers.
>
> (Fish, *Surprised by Sin*, 2nd ed., 1998, xiv, 340)

In *Self-Consuming Artifacts: The Experience of Seventeenth Century Literature*, Fish describes what he calls the *dialectical* text as "disturbing, for it requires of its readers a searching and rigorous scrutiny of everything they believe in and live by" (Fish, *Self-Consuming Artifacts*, 1972, rpt. 1994, 1). *Paradise Lost* is a dialectical text in educating the reader by disturbance, cognitive dissonance, and hermeneutical crises. With obscure allusions, geographical references, and theological asides, Milton's erudite narrator is schooling the reader in classical and contemporary knowledge, developing in the reader what Sheridan Blau calls the performative habits of "highly literate readers" (Blau, "Performative Literacy," 2003, 18–22). The reader is to the narrator of *Paradise Lost* what Adam is to Raphael: always in the posture of the learner, eager but dependent, curious but cautious, learning how to learn. What makes the audience of *Paradise Lost* "fit" (7.31) is that it has learned to read the poem with the wisdom and continence that Milton praises in *Areopagitica*. As a dialectical text, *Paradise Lost* invites critical engagement with its central claims regarding theodicy, also in questioning the justness of God in the catastrophic changes to creation after the fall of Adam and Eve.

The Nature of Creation

The earliest reference in *Paradise Lost* to the material cause of creation comes in Book 2, when the narrator describes the unformed elements through which Satan journeys on his way from hell to earth. Typical of Renaissance accounts of creation, such as Guillaume de Salluste Du Bartas's *La Semaine ou Création du Monde* (1578), Milton draws on classical and Christian traditions in his cosmogeny, particularly in the notion of Chaos as the personification of the primordial matter of creation. Du Bartas calls the raw materials of creation a "dull Heap of undigested stuff" (Du Bartas, *Du Bartas his deuine weekes*, Sylvester trans., 1608, C5^{r}), quoting Ovid's description of Chaos in Book 1 of *Metamorphoses*. Here is George Sandys's 1621 translation of the passage in Ovid:

> The Sea, the Earth, al-covering Heaven unfram'd,
> One face had Nature, which they Chaos nam'd:
> An undigested lump; a barren load,
> Where jarring seeds of things ill-joyn'd aboad.
> No Titan yet the World with light adorns;
> Nor waxing Phoebe fill'd her wained horns.
> (*Ovid's Metamorphosis*, Sandys trans., 1621, A1^{r})

The Geneva Bible uses a similar phrase, "As a rude lumpe," in its marginal gloss at Genesis 1:2 ("And the earth was without forme and voyde," A1^{r}). Henry Cockeram's *The English Dictionary* (Cockeram, *English Dictionary*, 11th ed., 1658) defines "Chaos" as "a confused and disordered heap, whereof poets feign the world, and all that therein, is to be made" (C1^{r}). Thomas Blount's *Glossographia, or A dictionary interpreting all such hard words* (1661) cites Ovid:

> *Chaos* (Gr.) a huge immense and formless mass, the rude and undigested first heap of natural elements; the world so called, before it was formed, as in Ovid.
>
> Unus erat toto naturae vultus in orbe,
> Quem dixere Chaos—
> (Blount, *Glossographia*, 1661, I1^{r})[5]

Dennis Danielson, in *Milton's Good God: A Study in Literary Theodicy*, neatly summarizes Milton's view as follows: "God in creation actualizes possible goods that exist in Chaos in a state of mere potentiality" (Danielson, *Milton's Good God*,

1982, 49). In this respect, Milton is mostly in line with the early modern classical and Christian accounts of the building blocks of creation.

Where Milton deviates significantly from these accounts is spelt out in his prose treatise, *De Doctrina Christiana,* which Milton left unpublished at his death. In Book 1, Chapter 7 ("On Creation"), Milton argues that the primordial matter of creation emanated from God:

> It seems that God could not have created this world out of nothing—"could not", not from the lack of power or omnipotence, but because there had to be something already in existence which by being acted upon might receive the almighty force of his efficacy. And so, since both Holy Scripture and reason itself suggest that all these things were not made from nothing but from matter, it necessarily follows that matter either always existed outside God or at some time originated from God. That matter should have always existed outside God . . . is not intelligible; nor, if it did not exist from eternity, is it any easier to understand where it eventually came from: There remains therefore only this idea, especially as scripture guides us to it, that all things came out of God.
>
> (Milton, *De Doctrina*, 2012, 289–291)

For Milton, not only is creation *ex deo* the only logical explanation, it is "a sign of [God's] supreme power and supreme bounteousness that such heterogeneous, multiform, and inexhaustible virtue should exist in God": "he does not, I say, enclose this heterogeneous and substantial virtue within himself but sends it out, propagates it, and extends it as far and in whatever way he himself wills" (*De Doctrina*, 293). Because the matter that became creation proceeded from God, creation is necessarily and essentially good:

> For indeed that original matter is not to be thought of as an evil or worthless thing, but as a good thing, a seed bank [*seminarium*] of every subsequent good. It was a substance, and derivable from no other source than from the fountain-head of all substance; at first unarranged [*indigesta*] and disorganized, but afterwards God arranged it and made it beautiful.
>
> (*De Doctrina*, 293)

The necessary consequence of this conclusion, according to Milton, "is that, since all things are not only from God but out of God, no created thing can perish into nothing" (*De Doctrina*, 295).

As a corollary of his heterodox position in *De Doctrina Christiana* regarding the material of creation as an *extension* from God, in *Paradise Lost* Milton holds a monist and materialist view that matter only differs from spirit in degree rather than in kind. When Adam asks the archangel Raphael about food, appetite, and angelic digestion, Raphael describes a process of reverse transubstantiation: "concoct, digest, assimilate,/And corporeal to incorporeal turn" (5.412–413).

> O Adam, one Almighty is, from whom
> All things proceed, and up to him return,

If not depraved from good, created all
Such to perfection, one first matter all,
Endued with various forms, various degrees
Of substance, and in things that live, of life;
But more refined, more spiritous, and pure,
As nearer to him placed or nearer tending
Each in their several active spheres assigned,
Till body up to spirit work, in bounds
Proportioned to each kind.
(5.469–479)

Angels are "ethereal substance" (6.330), who experience the same physical pleasure that humans do, but to a greater degree:

Whatever pure thou in the body enjoy'st,
(And pure thou wert created) we enjoy
In eminence . . .
(8.622–624)

Because the sexual attraction that Adam feels for Eve is rational and temperate, his physical love may be different *in mode* from spiritual life but not *in substance*: it "is the scale/By which to Heav'nly love thou may'st ascend" (8.591–592). As Milton notes in *De Doctrina Christiana*, "anything spiritual can be made out of a body" (295). Chaos was the unformed matter that, before time, proceeded out of God the Father and by which the Son and Spirit created the universe. The material universe, then, is a lower, physical expression of the highest, spiritual perfection that is God. Regina Schwartz, in *Remembering & Repeating: On Milton's Theology and Poetics*, concludes: "An evil chaos would indict, rather than justify a God from whom all proceeds, accusing him of fashioning a universe rotten to its very core . . . he must deliberately and explicitly assert the goodness of first matter" (Schwartz, *Remembering*, 1993, 8–9).

The Environmental Cost of the Fall

Milton follows the biblical text in Genesis when God the Father pronounces the penalty for sin on Adam and Eve. To Adam, God says:

Cursed is the ground for thy sake, thou in sorrow
Shalt eat thereof all the days of thy life;
Thorns also and thistles it shall bring thee forth
Unbid, and thou shalt eat th' herb of the field,
In the sweat of thy face shalt thou eat bread,
Till thou return unto the ground . . .
(10.201–206)

Cursing the ground is metonymic for all the changes to creation subsequent to the Fall that are described in those 60-plus lines of Book 10 quoted in the introduction earlier: "These changes in the heav'ns, though slow, produced/Like change on sea and land" (10.692–692). Milton's description of changes to creation echoes Renaissance theories regarding meteorology and cosmology, with the narrator admitting epistemic uncertainty about the best explanation for the altered cosmos:

> Some say he [God] bid his angels turn askance
> The poles of Earth twice ten degrees and more
> From the sun's axle; . . .
> . . . some say the sun
> Was bid turn reins from th' equinoctial road
> Like distant breadth to Taurus with the sev'n
> Atlantic Sisters, and the Spartan Twins
> Up to the Tropic Crab; . . .
> (10.668–670, 671–675)

Just as the narrator (through Raphael) was coy about choosing between the Ptolemaic or Copernican worldviews earlier in *Paradise Lost* (8.114–178), so the narrator does not settle the question about the cosmos after the Fall. What matters, though, is that the solar alteration occurs "At that tasted fruit" (10.687). The preposition "at" here is both causal and temporal: because the forbidden fruit was tasted *and* in that very instant, the sun turned from its original course. As Nigel Smith succinctly puts it: "The Fall is an ecological disaster" ("Paradise Lost and Heresy," *The Oxford Handbook of Milton*, 2009, 516).

From celestial to terrestrial, Milton describes the disastrous effects of climate change "on sea and land" (10.693). Eruption, vapor, mist, and exhalation are "hot,/ Corrupt and pestilent" (10.694–695):

> now from the north
> Of Norumbega, and the Samoed shore
> Bursting their brazen dungeon, armed with ice
> And snow and hail and stormy gust and flaw,
> Boreas and Caecias and Argestes loud
> And Thrascias rend the woods and seas upturn; . . .
> (10.695–700)

The military imagery, echoed in the harsh consonants and alliteration and amplified by the run of "and" conjunctions (called *polysyndeton* in rhetoric), heightens the sense of creation in conflict. The "Outrage from lifeless things" (10.707) is matched by "fierce antipathy" as "Beast now with beast gan war, and fowl with fowl,/And fish with fish" (10.709–711). Herbivores become carnivores, and humanity's fond friendship with creatures turns to fearful flight.

> nature first gave signs, impressed
> On bird, beast, air, air suddenly eclipsed
> After short blush of morn; nigh in her sight

The bird of Jove, stooped from his airy tow'r,
Two birds of gayest plume before him drove:
Down from a hill the beast that reigns in woods,
First hunter then, pursued a gentle brace,
Goodliest of all the forest, hart and hind;
Direct to th'eastern gate was bent their flight.
(11.182–190)

Adam recognizes that "Heav'n by these mute signs in nature shows/Forerunners of [God's] purpose" (11.194–195). Consistent with the recurring metamorphosis motif in *Paradise Lost*, as in the grotesque transformation of Adam and Eve's love into lust, perpetual change characterizes the cursed creation after the Fall.

Theodicy and Anthropogenic Change

When debating with himself whether it is better to live or die—his "to be, or not to be" soliloquy—Adam uses the phrase "sufficient penalty" (10.753) in questioning the further punishment on future generations for his transgression. The poem has demonstrated in the various uses of the word "sufficient" that it implies a completeness and efficacy commensurate with its purpose or object: "sufficient to have stood" (3.99), "thanks sufficient" (8.5), and "sufficient that thy prayers are heard" (11.252). "Sufficient" also implies enough to satisfy a requirement, particularly a debt or payment, as in Beelzebub's words about the purpose of the suffering in hell of the rebel angels:

But what if he our conqueror . . .
. . . .
Have left us this our spirit and strength entire
Strongly to suffer and support our pains,
That we may so suffice his vengeful ire . . . ?
(1.143, 146–148)

The biblical notion of Christ's propitiation as satisfying the required penalty is also invoked at key moments of the theodicy of *Paradise Lost*:

He, with his whole posterity, must die,
Die he or justice must; unless for him
Some other able, and as willing, pay
The rigid satisfaction, death for death.
(3.209–212)

Shortly after God the Father makes this pronouncement, the Son of God offers himself as a sufficient substitute: "So Man, as is most just,/Shall satisfy for Man, be judged and die" (3. 294–295).

What makes the death penalty for Adam and Eve and the life-substitution by the Son "sufficient" is two-fold. First, Adam and Eve were created capable and free to make rational choices, and are, therefore, responsible for the consequences of breaking what Adam blithely called that “One easy prohibition” (4.433). Milton had earlier emphasized Adam’s ability in *Areopagitica*:

> Many there be that complain of divine Providence for suffering *Adam* to transgress, foolish tongues! when God gave him reason, he gave him freedom to choose, for reason is but choosing; . . .
>
> (Milton, *Areopagitica, Prose Works*, 1959, 527)

Milton’s retelling of the Fall adds to the biblical account several ways by which God forewarns and prepares Adam and Eve for Satan’s attempt to destroy them and the new creation. Second, as Milton dramatizes in *Paradise Regained*, the Son of God descends into human form and resists Satan’s temptations in the wilderness, thereby fulfilling the requirement as substitute for Adam’s penalty. Both penalty and payment are “sufficient” to satisfy God’s just judgement.

After the Fall, Adam mentions to Eve that he effectively dodged a bullet when God cursed the ground rather than him: “on me the curse aslope/Glanced on the ground” (10.1053–1054). The Miltonic reader might come to the same ambivalent conclusion about God’s punishment of creation that Adam also does in parsing out his own sentence: “Inexplicable/Thy justice seems” (10.754–755). Can the existential changes to the natural world, consequent to the fall of Adam and Eve, be justified within the moral and ontological framework of the poem? In terms of the moral qualification so that the penalty is properly “sufficient,” creation itself must have agency: it must be shown to be capable if it is to be culpable. That creation cannot meet the criterion of being held morally responsible is best understood in terms of its relative ontological status. First, as Milton established in *De Doctrina Christiana* to his satisfaction on the basis of Holy Scripture and reason, because all matter is out of God (*omnia ex deo*), creation is itself good. God’s pronouncement of goodness upon his creation in Genesis (“and God saw that it was good”) denotes the aesthetic, functional, and moral quality of created matter. Matter is essentially good. Second, Milton’s monist ontology—that created things have one degree or other of spiritual corporality—considers everything below humanity not to have reason. Creation’s “various forms, various degrees/Of substance” (5.473–474) occupy their appointed places on a scale, with angelic beings more spiritual and rational than human beings, who are distinct from sentient and non-sentient created things by reason of their reason. Consequently, given their place on Milton’s ontological spectrum, animals and “lifeless things” (10.707) are incapable of rational choice and, therefore, are not responsible. The penalty on creation does not meet the standard of theodicy that the poem has established for angels and humans. There may be a valid theological justification for Adam’s curse falling “aslant” onto creation, but not on the basis of the ontological and moral frameworks established by Milton in *On Christian Doctrine* and *Paradise Lost*.

Performative Literacy

The question “Is God justified in punishing creation for the sin of Adam and Eve?” is deliberately provocative, particularly in the context of a Christian undergraduate classroom. Revealed things are ours to explore, but do not the "secret things belong unto the Lord our God" (Deut. 29:29)? And does *Paradise Lost* itself not teach the reader that transgressing the limits of inquiry has catastrophic results? Raphael's narration in Books 5 and 6 of Lucifer's rebellion and the war in heaven serves as a "dire example to beware/Apostasy" (7.42–43):

> He [Adam] with his consorted Eve
> The story heard attentive, and was filled
> With admiration, and deep muse to hear
> Of things so high and strange . . .
> . . . .
>
> Whence Adam soon repealed
> The doubts that in his heart arose: and now
> Led on, yet sinless, with desire to know
> What nearer might concern him, how this world
> Of heav'n and earth conspicuous first began,
> When, and whereof created, for what cause,
> What within Eden or without was done
> Before his memory, as one whose drouth
> Yet scarce allayed still eyes the current stream,
> Whose liquid murmur heard new thirst excites,
> Proceeded thus to ask his Heav'nly guest . . .
> (7.50–53, 59–69)

The epic simile in this passage suggests that Adam's "desire to know" is an unslakable thirst, one that "new thirst excites." Yet this is precisely the sort of boundary-testing of the responsible—potentially heretical—reader that Milton idealizes in *Areopagitica* and implies in *Paradise Lost.* Sheridan Blau calls such critical practice "performative literacy":

> Performative literacy can be identified as an enabling knowledge—knowledge that enables readers to activate and use all the other forms of knowledge that are required for the exercise of anything like a critical or disciplined literacy. It also represents a set of literate practices without which readers cannot continue to grow in knowledge and literary competence through their reading experience. This enabling form of knowledge—performative literacy—is essential to functioning as a fully enfranchised reader in 21st century schools.
> (Blau, "Performative Literacy," 2003, 19)

Enfranchisement is one of the central themes in Milton's prose work, whether he is writing about government, church, or marriage and divorce, and enfranchisement is a key indicator of an educated reader.

The provocative question about theodicy and postlapsarian climate change in the undergraduate classroom is first *centripetal* in that it draws the student into the text of *Paradise Lost.* To answer the question, students must first discover what constitutes "sufficient penalty" in terms of the theodicy that Milton is proposing.

As the opening invocation announced, this is the central purpose of the epic. Second, students must describe the characteristics or nature of creation in *Paradise Lost*, perhaps in reference to Book 1, Chapter 7, of *De Doctrina Christiana.* By describing the ontology and cosmology of *Paradise Lost*, students will begin to understand precisely what sort of world Milton is imagining, what sort of paradise has been lost. The question is also *centrifugal* in that it drives students to compare Milton's cosmos in *Paradise Lost* with other seventeenth-century accounts of Chaos and creation, specifically in relation to the New Science and the proceedings of the Royal Society of London. The question also invites students to consider ideologically inflected or confessionally oriented readings of *Paradise Lost*, such as William Empson's classic *Milton's God* (1961) or C.S. Lewis's *Preface to Paradise Lost* (1960) or the "New Milton Criticism."[6] Furthermore, the question of God's justness and the state of the fallen world leads students to probe the nexus of divine providence, eternal decree, eschatology, and human responsibility for the state of the planet. What is the theological or philosophical basis for environmental rights? Does creation have rights? How should we advocate for groaning creation (Romans 8:22)? What other categories or criteria would one use to justify the ways of God to creation, to the material world?

Such an active reading of *Paradise Lost* lends itself well to the liberal arts and sciences college classroom. Upper-level undergraduate students are historically engaged with environmental issues and are typically expecting to have their assumptions, preconceptions, biases, and prejudices challenged in their course work. Within the liberal arts and sciences framework, a literature course in which students encounter *Paradise Lost* is particularly attuned to the affective domain that encourages reflective reading. The sort of critical reading of *Paradise Lost* that provocative questions produce serves as a heuristic in the undergraduate classroom because it involves parsing the text's environmental grammar and posing the moral and ethical challenge of anthropogenic climate change. Rather than reading *Paradise Lost* as an apology for Enlightenment attitudes towards science, technology, and environment, an ecocritical approach allows the epic itself to become a way of probing the dominant narrative, of developing practical critical skills, and of encouraging divergent thinking about paradises lost and reclaimed.

Notes

1 See Leonard, *The Value of Milton*, 2016, 63–71; Edwards, "The 'World' of *Paradise Lost*," 2009, 496–509; Poole, "Creating a Universe," *Making of Paradise Lost*, 2017, 172–189; Danielson, "Multiverse, Chaos, Cosmos," *Paradise Lost and the Cosmological Revolution*, 2014, 27–50.

2 See, among others, Hiltner, *Milton and Ecology*, 2003; McColley, "Milton and Ecology," 2003; McColley, *Poetry and Ecology*, 2007; Watson, *Back to Nature*, 2008.

3 On the topic of Creation Care, see, for instance, materials prepared by the Lausanne Movement: https://lausanne.org/networks/issues/creation-care.

4 Raymond, "The Literature of Controversy," in *A Companion to Milton*, 204, writes: "*Areopagitica* imagined a world of dynamic readers, wrestling with material books, as well as with the substance of their words. These are the circumstances into which Milton pitched all his subsequent writings, both prose and poetry."

5 That the Chaos tradition continued into the Enlightenment and beyond is reflected in the opening movement of Josef Haydn's *Die Schöpfung*, 1798: "Die Vorstellung des Chaos."
6 See Herman and Sauer, *The New Milton Criticism*, 2012, 1–22.

References

The Bible: Translated According to the Ebrew and Greeke, 1611. 2nd ed. www.proquest.com/docview/2264238751.

Blau, Sheridan. "Performative Literacy: The Habits of Mind of Highly Literate Readers." *Voices from the Middle* 10, no. 3 (2003): 18–22.

Blount, Thomas. *Glossographia, or, A Dictionary Interpreting All Such Hard Words of Whatsoever Language, Now Used in Our Refined English Tongue . . .* London: Printed by Tho. Newcombe for George Sawbridge, 1661. www.proquest.com/docview/2264217449.

Cockeram, Henry. *The English Dictionary, or, An Interpreter of Hard English Words*. 11th ed. London: Printed by A.M. and are to be sold by Andrew Crooke, 1658. www.proquest.com/docview/2264214766.

Corns, Thomas N., ed. *A Companion to Milton*. Blackwell Companions to Literature and Culture. Oxford: Blackwell, 2003.

Danielson, Dennis. *Milton's Good God: A Study in Literary Theodicy*. Cambridge: Cambridge University Press, 1982.

———. *Paradise Lost and the Cosmological Revolution*. Cambridge: Cambridge University Press, 2014.

Du Bartas, Guillaume de Salluste. *Du Bartas His Deuine Weekes and Workes Translated*. Translated by Joshua Sylvester. London: Humphrey Lownes, 1608. *Early English Books Online*, 2011. http://name.umdl.umich.edu/A11395.0001.001.

Edwards, Karen L. "The 'World' of *Paradise Lost*." In *The Oxford Handbook of Milton*, 496–509. Oxford: Oxford University Press, 2009.

Empson, William. *Milton's God*. London: Chatto & Windus, 1961.

Fish, Stanley Eugene. *Self-Consuming Artifacts: The Experience of Seventeenth Century Literature*. Berkeley: University of California Press, 1972. Rev. ed. Pittsburgh: Duquesne University Press, 1994. All page numbers refer to the 1994 edition.

———. *Surprised by Sin: The Reader in Paradise Lost*. Cambridge, MA: Harvard University Press, 1967. 2nd ed. Cambridge, MA: Harvard University Press, 1998. All page numbers refer to the 1998 edition.

Herman, Peter C. and Elizabeth Sauer, eds. *The New Milton Criticism*. Cambridge: Cambridge University Press, 2012.

Hiltner, Ken. *Milton and Ecology*. Cambridge: Cambridge University Press, 2003.

Leonard, John. *The Value of Milton*. Oxford: Oxford University Press, 2016.

Lewis, C. S. *A Preface to Paradise Lost*. Oxford: Oxford University Press, 1960. https://archive.org/details/prefacetoparadis0000lewi (1959).

McColley, Diane Kelsey. "Milton and Ecology." In *A Companion to Milton*, edited by Thomas N. Corns, 157–173. Oxford: Blackwell, 2003.

———. *Poetry and Ecology in the Age of Milton and Marvell*. London: Ashgate, 2007.

McDowell, Nicholas and Nigel Smith, eds. *The Oxford Handbook of Milton*. Oxford: Oxford University Press, 2009.

Milton, John. *Areopagitica. 1644. Complete Prose Works of John Milton: Vol. 2*. Edited by Ernest Sirluck, 480–570. New Haven: Yale University Press, 1959.

———. *De Doctrina Christiana*. Vol. 1. John K. Hale and J. Donald Cullington, eds. *The Complete Works of John Milton*. Vol. 8. Oxford: Oxford University Press, 2012.

———. *Paradise Lost*. 1674. Edited by Stephen B. Dobranski. New York: W.W. Norton, 2022.

Ovid. *The First Five Bookes of Ovids Metamorphosis*. Translated by George Sandys. London: W.B., 1621. www.proquest.com/docview/2240937321.

Poole, William. *Milton and the Making of Paradise Lost*. Oxford: Oxford University Press, 2017.

Raymond, Joad. "The Literature of Controversy." In *A Companion to Milton*, edited by Thomas N. Corns, 191–210. Oxford: Blackwell, 2003.

Schwartz, Regina. *Remembering & Repeating: On Milton's Theology and Poetics*. Chicago: Chicago University Press, 1993.

Strong, James and John R Kohlenberger. *The New Strong's Expanded Exhaustive Concordance of the Bible*. Nashville: Thomas Nelson, 2010.

Watson, Robert N. *Back to Nature: The Green and the Real in the Late Renaissance*. Pittsburgh: University of Pennsylvania, 2008.

7 Early Modern Reformed Theology and Nonhuman Animals

Katherine M. Quinsey

Recent work in early modern animal studies has helped reveal some of the dualities that mark the seventeenth century as a watershed moment in the Western understanding of what it was to be human, and how humans related to nonhuman beings. As medieval and classical traditions met and informed both new science and older scepticism,[1] so animal sentience and intelligence were debated, species divisions on basis of the spirit-matter distinction were challenged, and human responsibilities for the nonhuman creation were reviewed. Nonhuman animals were beginning to be seen not only as symbols in a human-centred narrative but as autonomous, sentient beings who existed in a complex relationship with humans.[2]

In England, much of this debate was played out in sermons, devotional literature, and religious tracts. As observed by Erika Fudge, there emerged in this time a distinction between a Calvinistic theology based on the assumption of the weakness of fallen human nature, holding that humanity's God-given dominion over nonhuman animals was lost in the Fall, and a Scholastic tradition influenced by classical Greek dualism, that maintained an absolute ontological distinction between animal and human nature, based on the immortal principle of reason implanted within humanity by God.[3] The writings of both Luther and Calvin are deeply imbued with references to nonhuman animals, both literal and metaphoric.[4] Reformed theology was characterized by a scepticism about human reason, seeing it not as Right Reason, the faculty by which humans apprehended the divine and were linked to God in their unfallen state, but as fallen, prideful intellectual effort.[5] Luther, who sees Adam and Eve as reflecting God's image in their "enlightened reason,"[6] describes human reason as a "beast," seeing it not as the faculty linking man to God but rather an impious human-centred arrogance: "this beast which is called reason . . . is the fountaine and headespring of all mischiefes. For reason feareth not God, it loueth not God, it trusteth not in God, but proudely contemneth him" (Luther, *Commentarie . . . vpon . . . Galathians*, 1575, 106).

More specifically, Reformed theology held that humanity's God-given "dominion" over animals in Eden was lost in their fall into sin, which broke this harmonious relationship. The Edenic dominion enjoyed by humans in the unfallen world was seen as relational rather than ontological; the *imago Dei* (image of God) entailed God's quality of love for otherness. For humans, it entailed the ability to appreciate and understand our fellow creatures, the spiritual faculty to see God's glory reflected

DOI: 10.4324/9781003366744-7

in their inherent natures, and the responsibility to care for them.[7] The creatures' primary function was to be fully themselves, showing forth God's love in their own being, and thus enhancing humans' relationship with God. This unfallen relationship was broken by human sin, and animals, like the nonhuman environment, became antagonistic to humans, both hostile and fearful.[8] Nonhuman nature, itself innocent and not capable of sinning, was inevitably affected by human sin, both in the original event and in ongoing fallen human behaviour. The dominion given man in the fallen world was seen as a debased version of the original, rooted in fear and violence; the covenant with Noah and with all the creatures in Genesis 9, in which humans dominate animals through fear, and meat-eating is seen as God's provision for humans, was a covenant for that world. In his commentary on this passage, Luther says in his piquant way that while Adam would have found it "an abomination to kill a little bird for food," God "sets Himself up as a butcher" for Noah to supply his kitchen (*Luther's Works*, 1958, 2.133, qtd in Clough, "Anxiety," 2009, 46). It was generally accepted that vegetarianism was God's original plan for humans and animals in Eden and that meat-eating was a necessary characteristic of the fallen world. The responsibility to care for animals was still seen as a religious duty, however; animals were God's creatures, so humans had a religious responsibility to treat them with care and kindness, even when they were to be eaten. Old Testament stories and Biblical laws were frequently quoted to support an ethic of care for animals.[9]

More broadly, God was seen as active and present in the entire nonhuman creation. David Clough quotes from a 1527 tract in which Luther "insists that the power of God 'must be essentially present at all places, even in the tiniest tree leaf'":

> [God] must be present in every single creature in its innermost and outermost being, on all sides, through and through, below and above, before and behind, so that nothing can be more truly present and within all creatures than God himself with his power . . .
>
> (*Luther's Works*, 1958, 37.57, 37.58, qtd in Clough, "Anxiety," 2009, 51)

In his commentary on Genesis, Luther holds the orthodox position that God remains present, continues to give life to, and to see goodness in the cosmos:

> For God did not thus create these things, designing to forsake them when created, but he approves them and loves them still. The great Creator by his divine agency still simultaneously stirs, moves and preserves, after his own God-like method all things which he hath made.
>
> (Luther, *Genesis Commentary*, 2016, 73–74)

Reformation theologians drew widely from the ancient trope of "the book of nature," seeing the physical universe as direct evidence of both God's power and love, as "marks" or "characters" or "footprints," from which humans needed to learn about the existence and nature of God, and as a rebuke to human pride. Luther says:

> Every bird therefore and every fish is nothing more or less than a word of divine grammar or language; . . . But these divine things are thus written and

> ought to be diligently observed, studied and known by us, that we might learn to admire and adore the power of the Divine Majesty; and that we might edify and strengthen our faith from all these marvellous creation works of God!
>
> (Luther, *Genesis Commentary*, 2016, 71–72)

Even in the Protestant tradition of *sola scriptura*, the book of nature had prior divine authority, being directly authored by God without human intervention.[10]

All creatures, even those not obviously useful to humans, were created for a purpose beyond human conception and to glorify God in their own individual natures. Pertinent to our contemporary crisis of species loss, the variety of creation, including those species not seemingly relevant or attractive to humans, was held to be particularly illustrative of God's glory, in addition to their vital role in an interdependent cosmos; their existence also curbed human pride, by reminding humans of the limitations of their perspectives.[11] As Calvin writes in his commentary on Job,

> . . . the great varietie that sheweth it felfe euerywhere ought too stirre vs vp and moue vs to seeke God, and too submit our selues too him when we haue founde him, yea euen with such obedience, that although we perceyue not the reason of his works . . . Seing then that there commeth such diuersitie before our eyes: God doeth therein shewe vs his prouidence the more apparantly, & we be so much the more conuicted.
>
> (Sermons . . . vpon the Book of Job, 1574, 780, 787)

Moreover, these are ancient concepts in orthodox church tradition, expressed in various patristic writings; Augustine, for example, declares the goodness of creation in its variety as a rebuke to human pride and an affirmation of the glory of the Creator (*City of God*, 1984, Book 12, chs. 4, 5, 475–477).

Thus, many of these ideas are not unique to Calvinist Protestantism, being embedded in long-standing Church doctrine. God's creation of the cosmos *ex nihilo* out of love, the inherent goodness of the creation, God's active presence in the physical and non-physical worlds ("all things visible and invisible"), God's love for creation and ongoing care for even the tiniest elements of it, God's love of multiplicity, and the creation being made to glorify God and to fulfil its own nature—while these concepts vary in emphasis and more or less consistently hold to the notion of human "dominion," they are central to orthodox Church teaching from its early establishment, being formulated in response to various early controversies that attempted to separate spirit from matter and God from creation.[12] It is worth noting that these enduring elements of orthodox Christianity are attuned to an appreciation of the natural environment and to theocentric (as opposed to anthropocentric) positions, in contrast to the various heresies they combatted, such as Gnosticism, which were often marked by extreme dualism, and denigration of the material world.

The theological positions outlined earlier were obviously shared by the Catholic church as well. As Schreiner points out, the differences between Catholics and Protestants did not centre on their views of the nonhuman creation (*Theater*, 1991, 2). Some writers see Reformed theology as animal-friendly, promoting a radical equality of persons that can incorporate the duty of kindness and justice to animals (Watson, "Protestant Animals," 2014), and focusing on human sinfulness, weakness, and the failure of the will rather than the primacy of reason (Fudge, *Perceiving Animals*, 2000, 38). Others see Catholic tradition as more animal-friendly, not only with a long tradition of saints closely associated with animals and the nonhuman environment (Kearns, foreword to *Divinanimality*, 2014, xii) but also with the deeper question of incarnational and Eucharistic theology as reflective of God's informing presence in the material world. Nonetheless, the area where both sides have broad agreement is in the view of the nonhuman creation and its essential goodness, together with human responsibility for animal welfare.[13] While classical Greek influence in medieval Scholasticism informed the doctrine of the primacy of reason, with the notion that animals were subject to humans and created for their use, it was still recognized that dominion meant "care" not exploitation and that nonhuman nature was peculiarly revelatory of God's goodness (Yamamoto, "Aquinas and Animals," 1998, 81; Fudge, *Brutal Reasoning*, 2006, 80; Berkman, "Thomistic Theology," 2009, 21-40). The teachings of the early Church Fathers are often cited—in particular Irenaeus, Chrysostom, Tertullian—in relation to a number of these issues: God's presence in creation, the book of nature, the nonhuman world in the Book of Job teaching humility before God, millennial prophecies of interspecies harmony, the nonhuman creation awaiting redemption in Romans 8, and humans' proleptic duty as stewards of God to care for the other creatures.

Finally, and engagingly, the possibility of future life for animals was held by a number of well-known and orthodox figures, from patristic times through early modernity—based on the eschatological promise of new heavens, new earth, in which the nonhuman creation would be renewed and restored, and on ideas of *telos* or purpose, that God creates nothing in vain. Preece and Fraser argue that this

> view was consistent with that of Arnobius and Origen in the early Christian years, John of the Cross in the sixteenth century, John Milton and the Cambridge Platonists (including Henry More) in the seventeenth, and repeated by the Reverend John Hildrop, the Reverend Richard Dean, and Leibniz in the eighteenth. Moreover, Bishop Joseph Butler, with Paley the most influential Anglican theologian of the eighteenth century, thought it a nigh certainty that animals had immortal souls while John Wesley, founder of Methodism, could see no reason why they would not.
>
> ("Status of Animals," 2000, 256)

Thomas points out that "in the later seventeenth century many otherwise orthodox clergy regarded the issue of animal immortality as entirely open," including such well-known figures as Samuel Clarke, while

> in the eighteenth century those who felt that animal salvation was at the very least a possibility included Bishop Butler, the clergyman William Whiston, the philosopher David Hartley and the writer Robert Wallace. Those who

> thought it highly probable or even a certainty included animal-lovers like John Hildrop and John Lawrence, Dissenters like Matthew Henry, and Methodists like Adam Clarke and John Wesley.
>
> (Thomas, *Man and the Natural World,* 1983, 140)

Much earlier, however, Luther appears to have been open to the idea as well, in the well-known instance from his Table Talk on the topic of the possibility of heaven for his much-loved dog Tölpel: "Certainly . . . Peter said that the last day would be the restitution of all things. God will create a new heaven and a new earth and new Tölpels with hide of gold and silver" (Luther, "Gleanings," 1957, 9, qtd in Clough, "Anxiety," 2009, 48).

In the seventeenth century, there emerge Anglican sermons specifically devoted to the themes of nonhuman animal status and human responsibility. A strong thread within them saw nonhuman animals as autonomous beings separate from human definition, created not merely for the service of man but to glorify God in their own natures, as "fellow creatures" with humans. Humans are accountable for their treatment of animals in this life, and nonhumans will have a place in the "new heavens, new earth" to come. Moreover, animals are part of a covenant community with humans under God. These threads can be traced through Anglican writers of both Calvinist and humanist inclinations: Joseph Hall, Thomas Draxe, John Rawlinson, Henry King, Edward Chaloner, Thomas Adams, Henry More, and Godfrey Goodman, for example. They continue into the eighteenth century, with an efflorescence of specifically Anglican writing around animal welfare occurring in the mid to late eighteenth century: John Hildrop, *Free Thoughts upon the Brute Creation* (1742); Robert Morris, *A Reasonable Plea for the Animal Creation* (1746); Richard Dean, *An Essay on the Future Life of Brutes* (1767); James Granger, *An Apology for the Brute Creation* (1772); and Humphrey Primatt, *The Duty of Mercy and the Sin of Cruelty to Brute Animals* (1776), followed by a number of works in the Romantic period linking animal welfare and human rights. (See Simon, *Animal Rights*, 2002, 43, for a brief listing.) The association of animal welfare with human social justice, specifically the abolition of slavery, is well known to be linked to Dissenting and evangelical Anglicans in the Romantic period through to the late nineteenth century, from William Cowper through Anna Aikin Barbauld and John Newton to Anna Sewell (the Quaker author of *Black Beauty*, 1877), but many of the themes for which those writers are noted—the emphasis on animal sentience, suffering, and their right to happiness; their autonomous identity and purpose before God; the responsibility of humans to care for them; the Edenic world as one of harmony between human and nonhuman—are well established in the previous century, and are rooted in Church orthodoxy.

Three examples from this rich vein of Anglican preaching will have to suffice. Joseph Hall states that human dominion is tempered by responsibility for animals' happiness and wellbeing; it is of such importance to our spiritual welfare that we will be called to account for it in the last Judgment:

> I doe there fore owe awe vnto God; mercie to the inferior creatures: knowing, that they are my fellowes, in respecte of Creation; whereas there is no

> proportion betwixt me and my Maker. . . . We little think it, but God shall call us to an account, for the unkind and cruell usages of his poor mute creatures: He hath made us Lords, not tyrants, owners, not tormenters: He that hath given us leave to kill them, for our use, hath not given us leave to abuse them, at our pleasure; they are so our drudges, that they are our fellowes by creation . . ."
>
> (Hall, *Works*, 1625, 11, 935)

Rather than separating humanity from animals, Hall groups them together as fellow creatures, not to be compared with the Creator, seeing no ontological separation between human and animal.[14] Godfrey Goodman's notable sermon *The Creatures Praysing God* (1622) also sees the creatures as belonging to a community with humans: "Then let vs conceiue that all the workes of God make one corporation, and are as members of one body, vnder one common gouernment" (Goodman, *Creatures Praysing God*, 1622, 4). His main thesis challenges the primacy of human reason, seeing both nonhuman sentient experience and human encounter with it as a means of grace, and as a model for the humility required in human experience of theological mysteries. Goodman sees animal expressiveness as having access to grace for both human and nonhuman, as

> the Creatures praise God in their voices and sounds, which in effect are their prayers; in their naturall law, which is their decalogue; in all their qualities and affections, both naturall, and as they are sometimes meanes and conduit-pipes of grace, whereby nature is sanctified.
>
> (Goodman, *Creatures Praysing God*, 1622, 1)

Their expression matches our own in weakness and imperfection, and both are valued by God:

> . . . for he accepts our imperfect prayers, and descends to our weake|nesse. Thus the stocks and the stones in their silence, and in their naturall properties; the beasts in their sounds and their cries, in their sence and in their motions, all serue to praise him: . . ."
>
> (2)

The nonhuman has expression of its own, challenging the primacy of speech and reason. He also suggests here that in this humble experience we can recover something of the original relationship between human and nonhuman in Eden, a "fellow-feeling" and "affection with" them as well as adoration of God's power evident in them:

> And though they seeme dumbe, yet in verity and truth they speake in their silence: for obiecting, or presenting themselues to our view, shewing their excellent nature, their rare and wonderfull properties, therein they speake their Maker. Man perceiuing and apprehending this, falls instantly to

> admiration, which is a kinde of naturall trance, wherein his speech failes him, as if he were for a time strucken dumbe together with the dumbe Creatures, to shew his fellow-feeling and symbolizing affection with them . . .
>
> (Goodman, *Creatures Praysing God*, 1622, 2)

Thomas Draxe, in *The Earnest of Our Inheritance* (1612), provides a powerful description not only of the renewed life of the nonhuman creation in the world to come but also of humanity's responsibility for them in this life, seeing their suffering in it as the direct result of human sin. According to Draxe, animals will be part of the new heavens, new earth at the coming of the Lord, a point he supports from Scripture and many of the Fathers and later writers (Draxe, *Earnest*, 1612, 8–9). In the new world, animals shall serve humans only through glorifying God, fulfilling their own nature:

> . . . their spirituall and principall ends, to witt, the setting forth of Gods glorie: the matter of mans delight, and the exercise of his meditation and thankfulnesse, may and shall continue and abide. These ends are the perfection of the creatures in which they shall rest.
>
> (10)

The use of the nonhuman for human purposes of food, clothing, and labour is a feature of the fallen world; in the renewed world to come, they will return to their principal and original purpose—to glorify God and to be delightful to humans, to be part of human experience of God's goodness. Draxe's description of the "corruption" and "frustration" of the creation (Romans 8: 20–22) combines the eschatological meaning of this passage with the quotidian experience of nonhuman animals, a "bondage" both spiritual and physical, metaphoric and literal, in which they suffer for, and from, humanity's sinful nature:

> . . . mans sinne hath so corrupted, enthralled, peruerted, and daily paineth the creature, that it, weary of this intollerable bondage doth in it kinde, make pittifull complaint vnto God, and groaneth for it deliuerance, and if God should open the mouth of it, as hee did the mouth of *Balaams asse,* it would expostulate and reason the case with *Man,* and say master, what haue I done, that thou hast thus corrupted mee; and tormented [68] mee so many times, am not I thy seruant which thou hast vsed and employed since the first time vntill this day?
>
> (Draxe, *Earnest*, 1612, 15–16)

As in Hall, animal expressiveness is emphasized as indicative of their sentience and autonomy, and their rights under divine justice, which compels us to treat them well. They are inherently good, and any imperfections we might perceive in them are the result of our own fall from grace:

> Let vs in no wise curse, banne, lame or misuse any of the poore Creatures, knowing that if there bee any defect or vntowardnes in their nature; or any

> want of duty and obseruance in them, towards vs, our sin hath beene and is the cause and occasion of it.
>
> (Draxe, *Earnest*, 1612, 26–27)

A recurring theme even from patristic and classical writings is the prospect of the nonhuman creatures witnessing against humans in the final judgment; in Draxe, more particularly, sins against animals are extended to sins against other humans and God:

> If we would not encrease the creatures groanes and *torments,* nor cause them (in it manner) to crie out and exclaime, yea and to preferre a bill of enditement vnto the iust iudge against vs; let vs then neuer (at least for the time to come) abuse, peruert and misapplie them to anie Idolatrous or superstitious vse or, to hurt and hinder (much lesse to ruinate and vndoe) our neighbour, or to serue our vnlawfull and inordinate lusts whether of pride and pleasure, gluttonie and drunkennesse, or of wrath and reuenge, &.
>
> (Draxe, *Earnest*, 1612, 21)

Finally, animals are seen as our neighbours and companions, as sharing community; moreover, in a remarkable anticipation of the way in which the cult of sensibility in the eighteenth century blurred species boundaries (see Quinsey, *Animals and Humans*, 2017, 3–5), they are deeply empathetic, reflecting Christian values of rejoicing with those that rejoice, and weeping with those that weep:

> They are (as it were) our sworne seruants, they are our confederates, allies and good friends & kind neighbours, and ready to doe vs any office and seruice, therefore they reioyce with and applaud vs, when it fareth well with vs: and doe condole, mourne and suffer with vs, when it goeth ill with vs, or when wee are any way hardly bestead.
>
> (Draxe, *Earnest*, 1612, 23)

My focus in this chapter is on a later seventeenth-century clergyman named Thomas Hodges, sometime rector of Souldern in Oxfordshire.[15] Born in Northamptonshire, he matriculated at St John's College, Cambridge in 1633, graduating BA and MA in 1638 and 1640, respectively. He held the Souldern rectorship under the Puritans and had it renewed under the monarchy in 1662. While Hodges had Dissenting sympathies,[16] he was orthodox and did not support the regicides (he was one of the signatories to the letter to the Reverend Ministers of the Gospel in London); nonetheless, he appears to have been well thought of by the Parliamentary party and the Commonwealth—they elected him to a fellowship chaplaincy at All Souls' College Oxon, although he never took it up and seems to have been a Cambridge man all his life. Blomfield says he supported the restoration of Church and State, and took his DD at Cambridge as soon as Restoration accomplished this (1661); he was made Rector of Souldern under the King but resigned the following year, presenting the advowson to his college (St John's Cantab) and requesting that the living be given to his friend William Twyne. He took the Oath of Supremacy as would be expected when he took on the position of chaplain and then master at the Wokingham Hospitium, a position he held until his death in 1687.

I will be considering primarily two sermons preached respectively at Oxford University and at the adjacent town of Brackley, published together in quarto in 1675 under the title *The Creatures Goodness as They Came out of God's Hands, and the Good Man's Mercy to the Brute Creatures Which God Hath Put Under His Feet*. These two sermons, which were preached at least twice before being published (Blomfield, *History*, 1893, 138), are notable for a robust and evangelically tinged Anglican orthodoxy, applied to the specific topic of animal welfare. They articulate a focused and theologically based sense of animal-human relations, which puts humans and animals on the basis of a shared creatureliness, while ascribing a sacramental significance to human responsibility. More notably, the language and theology of Hodges's sermons vividly express a new sense of animal identity, sentience, and status, in a manner that anticipates not only the language of late eighteenth-century humanitarianism and its strong links with Dissent but also the notion of animals as having legal rights, almost 150 years before national animal welfare legislation was first enacted in Britain.[17] These two sermons bring together sharply notions of animal status that are both emergent and traditional in this time; Hodges articulates these themes in a unique and forward-looking way, reflecting an important strand in Anglican thinking that incorporates both the scepticism of Plutarch and the teachings of St. Paul. As Morillo points out, the efflorescence of eighteenth-century religious writing on these themes does not occur until after several decades of relative silence from the pulpit (*Rise of Animals*, 2017, 79), though Alexander Pope's Plutarch-influenced *Guardian* 61 on animal rights (1713) addresses them at length (Quinsey, "Little Lives," 2017, 152–155).

The first sermon outlines the theological justification of the status of nonhuman animals, taking for its text the opening verses of Genesis 1, in which God saw his creation, and declared that it was good. The second, taking for its text Proverbs 12:10 ("The righteous man regards the life of his beast"), develops the practical application of this theology in human and nonhuman life. In his introductory remarks, Hodges justifies devoting two sermons to the topic of animals: God speaks to us directly through nonhuman nature, he says, "and truly where God gives a Text, and preaches on it himself, his Ministers have a sufficient warrant to make a Sermon" (Hodges, *Creatures Goodness*, 1675, A2r). God uses the nonhuman animals to teach Job humility, and Solomon's wisdom does not consist of "Metaphysical Speculations and Notions" but of specific everyday examples—a line of argument Hodges shares with Luther and Calvin. The purpose of the second sermon is equally located in the quotidian world of humans and nonhumans, addressing abuse of animals in the here and now, while placing those actions in a theological light:

> As for my choice of the subject of the second Sermon and my Discourse thereon; my apology is, if it need one, that it hath often grieved my Soul, to see how the poor bruit Beasts have been used, or abused rather, by their

> inhumane, merciless, absurd, & unreasonable cruel Masters, *who having no understanding became worse than the Beast that perisheth.*
>
> (Hodges, *Creatures Goodness*, 1675, A2r)

In losing the gift of compassion and understanding, which will be key elements in defining "humanity" in the following century, these men have lost the key theological concept of *imago Dei*, God's divine quality in humans, and become worse than beasts.

Hodges opens with a vivid reformulation of the medieval "Book of Nature" tradition: we are

> to study the Book of Nature well, . . . and therein to see the invisible things of God, even his eternal Power and Godhead, which being of themselves not seen, yet are manifest by the things that do appear. . . . In the Universe, as in a fair Book and very legible Characters, you may read, yea you may run and read, the Name and Nature of God; and by reading and studying this Book only, you may prove good Scholars in Divinity.
>
> (Hodges, *Creatures Goodness*, 1675, 3)

The ineffable is understood through the physical created world, the unseen through the seen, the Book of Nature predating the written word—a notion that, as Harrison points out, was widely received, as far back as Tertullian (*Bible, Protestantism*, 1998, 194). Living creatures are inherently good, coming directly from God, and specifically revealing of God's own nature; they are "drops of that Ocean of goodness that is in God . . . so many Letters, out of which men might spell the Name of God Almighty, *Psal.* 19" (Hodges, *Creatures Goodness*, 1675, 6). Hodges brings together a number of specific images from this ancient metaphor, which was used by a wide array of early modern writers, and which was gaining renewed ascendency in this time. A notable proponent was the hermeticist John Webster:

> Many do superficially and by way of *Analogy* (as they term it) acknowledge the Macrocosm to be the great unsealed book of God, and every creature as a Capital letter or character, and all put together make up that one word or sentence of his immense wisdome, glory and power; . . . Alas! we all study, and read too much upon the dead paper idolls of creaturely-invented letters, but do not, nor cannot read the legible characters that are onely written and impressed by the finger of the Almighty; . . .
>
> (Webster, *Academiarum Examen*, 1654, 28)

Hodges portrays animals as almost in a mediatory role, as models of service and humility, and as evidence of God's goodness in their design, thus a means by which humans can know and ascend to God.

> Yea from these Creatures, as from so many Teachers, men might learn many profitable Lessons, touching service and obedience to their Maker. The Creatures

> proclaim to us God's goodness, and glory: and . . . the homage and duty, which we owe unto God our Creator. The Creatures are a kind of Ladder, whereby we may always ascend up unto God. The World is *Templum Dei*, God's Temple: and *in this Temple doth every one* of his Creatures *speak of his honour*.
>
> (Hodges, *Creatures Goodness*, 1675, 6–7)

These ideas are certainly propounded in Goodman's Creatures *Praysing God*, and the image of creation as a ladder of revelation forms the title metaphor of Bellarmine's well-known answer to atheism from the argument of design, *Iacob's ladder consisting of fifteene degrees or ascents to the knowledge of God by the consideration of his creatures and attributes* (1638). Hodges uses the image similarly here, to support the theistic argument from design that was becoming dominant at this time but also to suggest a kind of spiritual progression, a sacramental significance to the creation whereby humans grow in understanding of the divine nature through contemplation of the nonhuman creation. As do Goodman, Chaloner, and Webster,[18] Hodges also suggests the autonomous expressiveness of the nonhuman creatures; they "proclaim" and "speak" of God in a world which is "God's Temple" not made by human hands. The voice or "cries" of creatures, their sentience and expressiveness, will also be a key theme in depicting their suffering at the hands of humans and in the discourse of animal rights in the century to follow.[19]

Hodges echoes Calvin in saying that the Holy Spirit often sends us to school to "th'unreasonable Creatures," to learn from their moral and behavioural qualities such as faithfulness, love, obedience, and self-sacrifice. Like Calvin, he draws on both classical fable and medieval bestiary tradition in saying that humanity should learn moral and social qualities from the animals; as is typical of fables, and as is characteristic of all the Scriptural allusions here, he combines the traditional symbolic qualities of animals with natural observation of their behaviour. Using phrasing similar to Calvin's in the commentaries on Job, Hodges quotes Biblical examples to show how humans can learn both moral and spiritual qualities from animals, while adding some popular elements:

> God would have us go to school to the Creatures, to learn many good Lessons from them. He would have the Sluggard go to the industrious *Ant; to consider her ways, and be wise.* Those who are ignorant of God their Lord . . . are reproved by the bruit Creatures; *For the Oxe knows his Owner, and the Ass his Masters Crib*. Those who do not, or cannot discern the times and seasons, even the times of their Visitation, the Prophet would that we go to *the Stork, the Crane, and the Swallow, for all these know their appointed seasons* [Proverbs 6:6; Isaiah 1:3; Jeremiah 8:7]. . . . How great is the love and faithfulness of Dogs! the meekness of Elephants! the shamefastness of the adulterous Lyoness! the chastity of the Turtle! the watchfulness of the Cock! the utility of the Sheep! 'Tis said of King Porus his Elephant, that he exposed his own life to save his Masters. How often does the Holy Ghost send us to school to the unreasonable Creatures!
>
> (Hodges, *Creatures Goodness*, 1675, 38)

The net effect is another challenge to the primacy of reason as a distinguishing human trait, as animals are seen to exemplify more purely the moral qualities of Christianity in their faithfulness and meekness; as Draxe observes, animals are exemplary in their uncomplaining suffering at the hands of fallen sinful humans (Draxe, *Earnest*, 1612, 27–30), a theme Hodges goes on to develop in his second sermon.

Nonhuman animals are made not merely for human service but to glorify God in the diversity of their own natures and behaviour; we see God's glory "1. From the various Natures of the Creatures./2. From the goodly Order of their creation, propagation, and working./3. From their great Usefulness, and serviceableness unto Man./4. From their extraordinary Obedience unto God, their Maker" (Hodges, *Creatures Goodness*, 1675, 7). In general, Hodges emphasizes the similarity between humans and nonhumans as created beings rather than ontological difference: "Come we to the Earth; our Bodies are made of it; we dwell therein; It brings forth Grass, and Fruits; and so, food for us, and the other Creatures; It is our Bed, to sleep in after death."[20] In a manner anticipating today's ecological perspectives, he emphasizes the limitations of human knowledge of animal nature: "God knows wherefore he made them [animals] all, and to what use they serve; although we do not" (Hodges, *Creatures Goodness*, 1675, 13). This ancient doctrine, that we do not know the full extent of the natural world, and that all creatures have a purpose even if it is not apparent to our limited perspective, speaks to ecological science today, as we grapple with our ignorance of the consequences of reducing or eliminating animal species. This is virtually a commonplace across a wide array of Catholic and Reformed traditions: Augustine, Bellarmine, Calvin, Chaloner, Henry More, and Peter Viret all state that the diversity in creation both reflects the glory of God and underscores the limitations of human perspectives—a Christian version of the ecological humility we so urgently require today. Hodges locates the goodness of this variety in New Testament teachings: "every Creature of God is good" (1 Timothy 4:4) and "All things shall work together for good, to them that love God, and are called according to his purpose" (Romans 3:8) (Hodges, *Creatures Goodness*, 1675, 6, 15). These claims of Christian salvation are re-applied by Hodges specifically to the animal-human relationship.

Hodges's view of the Fall and its effect on animal-human relations is firmly rooted in Reformation theology. Humans were subject to original sin, but animals retained their original innocence, described by him as their "obedience to God." Animals at this time were still seen as having moral agency, while at the same time instinct was the law of God written in the animal's nature. Notably, Hodges does not invoke the usual argument of human exceptionalism, which is that only man had reason and the freedom of will, to sin; rather, he emphasizes the inherent goodness of nonhuman animals.

> It may be said of all God's Creatures, as they were made at first by him, from the highest Angel in Heaven to Man on Earth, and the meanest Worm in or under the Earth; that, God commanded and they were created, and they rebelled not against his Word. . . . [and although the angels and Adam rebelled] . . . yet the residue of the Creatures, even the whole Creation . . . keep the Law that their Maker set them, and are still obedient unto his Word. . . .[21]

Reformed theology focuses on the infected will rather than on the primacy of reason, and on the perversion of human nature that resulted, which distorted the relationship between humanity and the nonhuman creation. This human-animal relationship is continually portrayed by Hodges in the light of lapsarian theology: he says later in the second sermon:

> We have sinned against God, which these Creatures never did. Man being in honour, understandeth not . . . but is compared to the Beast that perisheth. 'Tis no Sin to be a Beast, but 'tis a great Sin to be like a Beast.
>
> (Hodges, *Creatures Goodness*, 1675, 39)

Humans lost their original status, defacing the image of God, and have sunk lower than beasts, losing the understanding that would connect them intimately to those beasts; a beast itself is a good thing and a creature of God, but a failed human has no place in the order of creation.

While the creatures are still good, they are now hostile to humans, or afraid of them.

> As for the inferior Creatures; Man's fall exceedingly crushed them; yea brought a curse on his own head, and theirs too: So that, now, they are not generally so pleasant, so useful, so serviceable, and obedient to us, as they would have been, if we had not sinned.
>
> (Hodges, *Creatures Goodness*, 1675, 15)

Although the creatures are not as good as they were at first, they "are stil good in respect of their essence, or being, which they have from God" (15). Harmfulness in animals is the result not of their nature but of human sin. "If any of God's Creatures, or Providences, prove evil and hurtful to us, let's blame our selves, let's blame our sins" (21). If there are noxious animals in the creation today, it is because of man's sin; as Hodges says, man was the rebel in the house, no wonder God began to pull it down around his ears; man was the briar and thorn in the garden, no wonder the earth brings these forth to him.[22] "Let's not blame, or find fault with God, for any of all his works, either of Creation, or Providence: for God hath made all things well, and he hath made every thing beautiful in his season . . ." (Hodges, *Creatures Goodness*, 1675, 21). Hodges is much more emphatic than some of his predecessors in repeating the point about the essential goodness of Creation even though the Fall has distorted the original relationship, focusing on the nonhuman creatures rather than on human sin, though the responsibility—and the sin—is humanity's. Animals still retain their essential goodness, though constrained by the situation in which they find themselves.

In both sermons, Hodges expounds in particular on Romans 8:20–22, on the idea that the created universe groans in bondage to sin while awaiting the glorious liberty of the children of God and will partake in their renewal—a passage that was often considered with respect to the future life of the nonhuman. In Hodges, animals' suffering of the present time is seen as the fruit of human sin and is translated directly as their current experience of human cruelty, which is then given a theological significance. On the text "*Rom.* 8. 20. *The Creature it self is made subject to vanity*," he explains that "there is a two-fold vanity to which the Creatures are now subject; 1. To be abused vainly, by sinful Men, to ryot, luxury, and excess; And 2. The Creatures are subject to corruption" (Hodge, *Creatures Goodness*, 1675, 15). "Ryot, luxury, and excess" could well refer to the cruelties practised through refined cookery (a common theme in animal welfare writing dating at least from Plutarch) as well as to the blood sports critiqued by the Puritans.[23] Man's abuse of the animal creation is a fruit of the Fall; the original relationship of loving care and obedience has been perverted into objectification of animals, seen here as twisted and unnatural. In response to the question, will the creatures be restored to their original goodness? that is, is there a future life and future redemption for nonhuman animals? Hodges tends to the conservative side, saying it is up to the reader to judge, though the Scripture suggests that "it is not improbable" that the nonhuman creation will be redeemed, but only through a fire that purifies and recreates all things:

> There are too, who think that the Creatures *waiting, groaning and longing to be delivered into the glorious liberty of the Sons of God*, Rom. 8. Doth hint some Restitution of the Creatures to their first good and sound state. Which whether it do or no, I leave it to you to judge. Only this I say, that if the visible World shall pass through a purgatory-fire at the day of Judgment, and if it shall be continued for the blessed Saints to contemplate God's goodness therein at first, and the glory of Christ the second *Adam*, who came to repair the ruins which the first made; that then it is not improbable, that the Creatures may the attain to their primitive goodness.
>
> (Hodges, *Creatures Goodness*, 1675, 17–18)

There were competing opinions on this at the time, but Hodges here echoes Calvin's thought that the matter of creation will be redeemed and restored to its original purpose.

Animal suffering is metaphoric for Christian suffering and redemption, made more literal in specific application to the practices of his own time; animals become almost Christ-like in suffering by, and for, our sins, undergoing the punishment that is merited by us:

> Let us consider, these Creatures were partakers of the curse of Mans Sin; by reason of this *they groan and travail in pain* their yoke is harder and their burden heavier: We should look on their sorrows and sufferings as occasioned, as merited by us, by our disobedience to our soveraign Lord. How do they spend their days in our service! end their lives in our service, for Good, for Physick; sometimes for a Sacrifice, and oftentimes even for a Pleasure, Delight, or Recreation! They suffer for us; . . . they suffer with us, as in War and other publick calamities; and let this suffice, let them never suffer unnecessarily from us. If we do, we make them exceeding miserable: the poor abused Beast hath no knowledge or foresight of death (as Men in misery have) to put an end ere long to his bondage and misery; nor any hopes

> of a Resurrection, or future regard for his sufferings (as good Men have) to comfort them. The Turks, I read, hold that the Beasts shall rise again; and although that is no part of my Creed; yet I dare say, if they should, they will rise in judgment against those that have cruelly used them here.
>
> (Hodges, *Creatures Goodness*, 1675, 37–38)

The orthodox exceptionalist argument that only humans have a sense of the future, or of a future state, is here carefully formulated and added to human responsibility for animals: they should be treated well in this life especially as they have no hope for another (another commonplace, echoed by Alexander Pope in his *Guardian* 61, 1714, 376), while the suggestion is also made that if animals were to partake in the resurrection they would also be on God's side in judgment of human sin—nonhuman creatures become almost proxy for God and God's judgment. Moreover, humans have a duty "towards the repair of the ruins of the Creation, to restore the Creatures to their primitive goodness and beauty" (Hodges, *Creatures Goodness*, 1675, 18) by their appropriate treatment of nonhuman animals. Thus, humane treatment of animals is placed in the eschatological context suggested by Romans 8:20.

The second sermon focuses more particularly on the animal-human relationship, emphasizing that animals have divinely ordained moral rights in this life, and together with humans, are part of a covenant relationship with God. Here Hodges parallels the experience of animals to that of the Israelites under Pharaoh, associates cruelty to animals with human arrogance towards God, and challenges the idea of animals as objects owned by humans. A similar theology linking animal abuse to human slavery will undergird the abolitionist movement a century later but also appears prototypically in the works of Thomas Tryon, published during Hodges's lifetime (see Simons, *Animal Rights*, 2002, 42; Thomas, *Man and the Natural World*, 1983, 171, 180). More specifically, respect for animal welfare is seen as the basis for divine laws of the Old Testament, part of covenant between God and Israel.

> Although God hath given us a Right to the bruit Creatures, and a Dominion over them; yet 'tis with subordination to himself and his Laws. . . . We are Lords over them, 'tis true; but we must not play the Tyrants over them. The *Israelites*, God's own People, had a good Title to their Beasts or Cattel, and yet he bound them to their good behaviour towards them by sundry Laws and Rules in reference to them.
>
> (Hodges, *Creatures Goodness*, 1675, 26)

He follows with examples such as not muzzling the threshing ox, and the Sabbath being for animals as well as humans (26–27, 34). He discourses at length about humane treatment of animals in respect of eating them, seen in Old Testament rules and laws, such as killing first and pouring out blood before eating, as well as before burnt sacrifice, and the rules about not seething kid in its mother's milk, with the comment:

> God most wise, most merciful would have his People not only abhominate all cruelty to the bruit Creatures, but even to abstain from all appearance of it. And that he might teach them mercy by his own example, that they might

> be merciful to the bruit Creatures as their Father in Heaven is merciful, He ordained them to be offered up upon his Altar, but not till they were first slain, and their blood in which their life is, was poured out, at the foot of it. God would have no Sacrifice no not of a Beast to be burnt alive upon his Altar.
>
> (Hodges, *Creatures Goodness*, 1675, 27)

In describing the behaviour of the righteous and the unrighteousness man, Hodges reviews the notion of human dominion: humans are both in a Godlike position of authority over animals and yet are also fellow creatures. They are only Godlike when they show God's qualities of compassion and care—a key concept in eighteenth-century humanitarianism, which linked the cult of sensibility with Calvinism. If the original animal-human relationship was perverted through the perversion of human nature in the Fall, redeemed Christian human nature should express itself as "humanity" in the sense of compassion rather than reason, and by which human care for animals directly parallels God's care for creation; in theological terms, Hodges defines the *imago Dei* both as reflecting the character of God and as doing God's work in a practical sense. The passage also emphasizes the "cry" of the creatures and their own sentience and expressiveness, which is paralleled in the human master's "cry" to God and empathy for his animals.

> . . . a good Man, who is the Child of God and Image of God, the work of his Heavenly Father he will do; he will be mild, gentle, merciful, not only to Brethren, Neighbours, and Strangers, but even to the bruit Creatures, so saith the Text, *A good Man, a righteous Man, is merciful to his Beast*. . . . He gives them their meat in due Season. . . . What God gives him for them, that they gather from his hand. If God send a Drought and the Fire burn up or devour the Pastures . . . so that the Beasts groan . . . the righteous Man, their Lord and Master, sympathizeth with them, is grieved for them . . . as God for his People, so he for his Cattel, in all their afflictions he is afflicted; and therefore when the Beasts cry unto him, 'tis not in the power of his hand to help them; what doth he but stretch out his arms and cry unto God for them?
>
> (Hodges, *Creatures Goodness*, 1675, 27)

Moreover, care for animals is a marker of obedience to God; in the context of God's punishments on Egypt in the Book of Exodus, Hodges remarks, "When God threatned a grievous hail upon *Egypt*, he commanded them to send and gather the Cattel that were in the Field, lest they should be slain with the hail; and the Scripture tells us, *He that feared the Word of the Lord amongst the Servants of* Pharaoh,[39] *made his Servants and his Cattel flee into the Houses, and he that regarded not the Word of the Lord, left his Servants and his Cattel in the Field*" (Hodges, *Creatures Goodness*, 1675, 28) and he sees it as a "grievous sin" in Pharaoh that he took no steps to protect the animals from the promised disasters (28).

The everyday reality of unkind treatment of animals is placed even more emphatically into an eschatological context by allusions to the parables on mercy to humans as a duty to God in the Gospels, underlining the shared creatureliness and household community between human and nonhuman, and giving animals almost Christ-like virtues reminiscent of the suffering Servant in Isaiah 53:[24]

> As God will require the life of a Man of the Beast that slayeth him, the Beast shall be put to death [Genesis 9:5, Exodus 21:29]. So will God call Men to accompt for the lives of their Beasts, if they be cruel unto them. . . . Let us consider, that we may deal merciful with these Creatures, that they are Fellow-creatures with us, yea our Fellow-servants to the great Lord of Heaven and Earth, as well as they are our Servants, and we Lords over them. And must give an account when our Lord Christ comes, if he find us beating our Fellow-servants, if he find us drinking with the Drunken, and presently falling upon these our Fellow-servants. . . . As they are our Servants, to consider that they are good Servants to us, who are too often bad Masters to them. If we will use them, they submit their necks to our yoke; if we abuse them they do not complain; In their mouth is no reproof. . . . these our Servants they are mute, dumb Creatures; they cannot plead their own cause, they return not a word for a blow, not a word for many blows, though without a cause, or a sufficient cause.
>
> (Hodges, *Creatures Goodness*, 1675, 37)

Hodges's description of the everyday cruelties practised upon animals by "unrighteous" men could come straight out of the Victorian humanitarian tradition exemplified in animal autobiographies such as *Black Beauty*. He shifts focus here to direct social observation, the need for human laws to protect animals, and the concept of animals as having rights under natural law; but he places it in the context of the groaning of creation in Romans 8, and the slavery of the Israelites in Egypt, so that the specific social phenomenon is given an eschatological and theological meaning. Like the nineteenth-century writers, he indicates that this kind of cruelty is built into the society, as standard social practice, and he suggests an eye-witness element; his text is now from life:

> How oft are these poor dumb Creatures beaten cruelly with many stripes, and that when they know not their Masters will, what it is they would have them do! How have I heard wicked Men curse and ban their Beasts! How have I seen them beat and abuse them without a cause; threatening (with *Balaam*) if there were a Sword in their hand to slay them! If these dumb Creatures should have their mouths opened as *Balaam*'s Asse's mouth was; how might they complain and say, Wherefore hast thou smitten me, it may be, not these three times only, but these threescore times! Although many Men live on the labour and sweat of their Beasts; yet they care not in their rage and fury to slay them, to spill their blood. Well may the whole Creation groan wayting to be delivered from this bondage unto corrupt and cruel Men. How many unreasonable and cruel Masters are there, which are like Pharao's Taskmasters to the Israelites, who required brick and gave them no straw, which demand of their Beasts to labour, to travel, to work, and allow them

> no meat, or not meat sufficient! How do Men load their Beasts, more than they are able to bear, and drive them faster than they are able well to go; and if they stumble, fall down, or stand still, how are these their Servants beaten, when the fault is in their cruel Taskmasters? I confess I have often wished, that some Law might be made for the case of these poor Beasts, even to bind their Masters to their good behaviour towards them.
>
> (Hodges, *Creatures Goodness*, 1675, 32–33)

He concludes with a classical dictum, commonplace by this time, that cruelty towards animals leads to cruelty towards humans, and mercy and kindness towards animals teaches us to be more merciful to our fellow humans. Violence breeds violence; even spectating violence hardens the heart and naturalizes cruelty; Hodges cites the example of ancient Rome:

> Even doubtless the cruel Spectacles at *Rome* did render the People less merciful . . . so he that looked on these bloody Spectacles with a desire that one should kill the other, and delighted therein when done, committed Murther against his Brother in his heart: . . .
>
> (Hodges, *Creatures Goodness*, 1675, 42)

This sentiment will be repeated by Locke in *Some Thoughts on Education* (1693), referred to by Pope in *Guardian* 61 (1713), savagely reiterated in Hogarth's *Progress of Cruelty* series in the 1740s, and will become a central thread in both children's literature and the poetry and polemic of the late eighteenth century:

> Doubtless as unto us, so for our sakes chiefly and ultimately was this written: *A good Man is merciful to his Beast, but the mercies of the wicked are cruel:* As if the Wise-man had said, A just or good Man is merciful to the Beasts; but a wicked Man is cruel and merciles to Mankind. And take this along with you; That, to use to shew mercy to our Beasts, is a way and means to make us merciful to our Brother and Neighbour.
>
> (Hodges, *Creatures Goodness*, 1675, 42)

Hodges's animals had to wait nearly 150 years for national legislation, but his sentiments directly anticipate the later debates that gave rise to both human rights and animal welfare legislation in the early nineteenth century. Writing in the latter half of the seventeenth century, Hodges draws upon the live terms of the early modern debate on the emerging status of animals, and in his Biblical and sacramental interpretation of the social phenomenon, anticipates the humanitarian movement of the next century.

Notes

1 English translations of Plutarch's *Morals*, which include his famous defences of animal intelligence, were published at both ends of the seventeenth century, and an expanded version of Plutarch's Gryllus dialogue, which reverses the animal-human role in the Circe myth, in 1549 (Raber, *Animal Bodies*, 2013, 1–4; Shannon, *Accommodated Animal*,

2013, 150ff.). Montaigne's *Essays* were available in translation in 1613 and in an opulent folio in 1685. The resurgence in scepticism coincided with both post-Reformation views of the weakness of human nature and the new science's radically expanded views of both cosmos and microcosmos, challenging earlier views of human centrality.

2 There are numerous recent studies of human understanding of the human-animal relationship as it evolved in this period, with particular attention to its nuances and contradictions, while at the same both scientific discovery and changing models of "humanity" blurred distinctions between human and nonhuman animals. See Thomas, *Man and the Natural World*, 1983; Harrison, *The Bible, Protestantism*, 1998; Agamben, *The Open*, 2004; Fudge, *Borders*, 1999; *Perceiving Animals*, 2000; *Brutal Reasoning*, 2006; Nash, *Wild Civility*, 2003; Steiner, *Anthropocentrism*, 2005; Shevelow, *For the Love of Animals*, 2008; Wolfe, *What Is Posthumanism*?, 2010; Lamb, "Sympathy," 2011; Wolloch, "Animals," 2012; Brown and Alkemeyer, "Rational Elephants," 2013; Shannon, *Accommodated Animal*, 2013; Raber, *Animal Bodies*, 2013; Cole, "Animal Studies," 2019.

3 Fudge, *Perceiving Animals*, 2000, 38. Watson emphatically links animal welfare discourse with Puritan religion and politics ("Protestant Animals," 2014).

4 Both, notably, rejected allegorical readings of animals in Scripture in favour of a literal interpretation, which meant that passages about such creatures as Leviathan or Balaam's ass were read as primarily about real animals rather than being merely symbolic of human issues.

5 During the early Enlightenment, an ontologically based Stoic concept of immortal reason that mirrored cosmic order and linked humanity to God was being replaced by our modern concept of reason as ratiocination, based on the evidence of the senses. This blurred the Aristotelian distinction between the "sensitive" and "rational" soul, which had been used to separate humans from animals—implications of which can be seen in Locke's model of experiential reasoning in the *Essay on Human Understanding* (1690).

6 Unfallen man "knows God and is the image of God, in [him] the similitude of the divine nature shines forth through his enlightened reason, through his justice and his wisdom" (*Luther's Works*, 1958, vol. 1, 66, qtd in Clough, "Anxiety," 2009, 44).

7 See Preece and Fraser, "Status of Animals," 2000, 246–249; Kemmerer, *Animals and World Religions*, 2011, 178–179.

8 Discussed by Schreiner, *Theater*, 1991; Ickert, "Luther and Animals," 1998; Fudge, *Perceiving Animals*, 2000; Clough, "Anxiety," 2009; Watson, "Protestant Animals," 2014. McLaughlin traces the history of this belief in the Church Fathers, as well as the belief in future restoration of this harmony in the eschaton—and the proleptic duty of kindness to animals in the interim ("Evidencing the Eschaton," 2011).

9 The ecological implications of Old Testament laws and the agricultural economy they reflect are well explored in Davis, *Scripture, Culture, and Agriculture*, 2009, and in Baukham's essays on "Jesus and Animals," *Animals on the Agenda*, 1998, 33–60.

10 Harrison, *Bible, Protestantism*, 1998, 194. As Harrison points out, this trope gained ascendancy during the seventeenth century, as new science produced vastly expanded concepts of the universe, together with the theocentric perspective, to remind humans of their limited perspective and to expand their wonder at God's infinite creative power. See also Thomas, *Man and the Natural World*, 1983, 166–167.

11 According to Peter A. Huff, "the Western intellectual tradition inherited by the sixteenth-century Reformers treated the diversity of life on earth as a standard category of the Christian theological enterprise, often speaking of non-human creatures as components in a "second book" of divine revelation" (Huff, "Calvin and the Beasts," 1999, 68).

12 See Schreiner, *Theater*, 1991, 1–2, 8, for an admirable summary of this point.

13 Thomas describes the consistency of the Church position on animal welfare from medieval through to Victorian times (Man and the Natural World, 1983, 154).

14 Cf. Calvin, *Sermons on . . . Job*, 1574, 786: "what proportionablenesse or nerenesse shall there be, if we make cōparison betwene God & his creatures?"

15 This Hodges is not to be confused with Thomas Hodges of the ODNB, a London Dissenting clergyman noted for somewhat heretical (and hermetical) opinions. Our Hodges

was a moderate Anglican of a "gentle Presbyterian" persuasion and a lifelong Cantabrigian. The most detailed biography of our Hodges can be found in the Blomfield's *History of the Present Deanery of Bicester, Oxon: Fritwell, Souldern*, 1893, 135–139, 155.

16 "He was a zealous Presbyterian, but of a gentler spirit than many of that sect. His coming to Souldern was at once marked by the introduction of the Hour Glass into the Parish Church, and, a change of service, wherein preaching was exalted above prayer." Qtd in Blomfield, *History*, 1893, 136.

17 Watson points out that laws restricting cock fighting and horse racing were instituted under the Puritans during the Interregnum; earlier laws against various cruel practices existed in Ireland (1635) and in Massachusetts (1641) ("Protestant Animals," 2014, 1115–1116, 1143 notes 20–22). The only essay paying any sustained attention to Thomas Hodges is in Morillo, *Rise of Animals*, 2017, ch. 3, 72–79.

18 *Six Sermons*, 238–239: "Aske the beasts and they shall teach thee; and the fowles of the aire, and they shall tell thee; speake to the earth, and it shall shew thee; or the fishes of the sea, and they shall declare vnto thee, who is ignorant of all these, but that the hand of the Lord hath made them. . . . All creatures are *Regij professores,* professors of that great King, all are his witnesses, all his preachers." See also 243–44: "In the booke of Nature we must thinke no page vnwritten on, wee must suppose euery creature, euen the basest to speake vnto vs; the starres of the firmament to crie out, and by their light to inuite vs to that eternall light which is aboue; the winds in the aiery regions to crie out, and admonish vs of the Spirit of the Lord which dwelleth in all things; the flouds and streames of running water to crie out, and summon vs to that crystall riuer and fountaine of liuing water which is in heauen; the earth when it trembles, and when its massie frame is shaken, to crie out, and put vs in mind of the ruine and dissolution of the world. And for as much as God in this world can be discerned by vs but in his hinder parts, that is, in his workes and his effects, lest they which he hath now left vnto vs as witnesses to enforme vs, hee one day produce against vs, as witnesses to condemne vs." John Webster describes the creation as speaking directly for God in its own secret language, now lost to humans: "so every creature understands and speaks the language of nature, but sinfull man who hath now lost, defac't and forgotten it" (*Academiarum*, 1654, 27).

19 Explored by Menely, *The Animal Claim*, 2015.

20 Hodges, *Creatures Goodness*, 1675, 13. He is echoing Luther here. Clough ("Anxiety," 2009, 49) finds it "striking" that Luther "repeatedly" recognizes similarities between humans and nonhuman animals: "Human beings share the life of land animals: they were created on the same day and had a 'common table' in the herbs and fruit of trees. (*Luther's Works*, 1958, 1.36) . . . . The beasts 'greatly resemble' human beings: They dwell together; they are fed together; they eat together; they receive their nourishment from the same materials; they sleep and rest among us. Therefore, if you take into account their way of life, their food, and their support, the similarity is great" (*Luther's Works*, 1958, 1.56). Hodges's description here draws from Luther but also looks ahead to Pope's description of the Edenic state of nature in *Essay on Man*, Epistle 3, lines 152–153, where "Man walk'd with beast, joint tenant of the shade;/The same his table, and the same his bed" (*Poems of Alexander Pope*, ed. John Butt, 1963, 530). This is clearly a common thread.

21 Hodges, *Creatures Goodness*, 1675, 14. Cf. Luther: "All animals live in contentment and serve God, loving and praising Him. Only the evil, villainous eye of man is never satisfied" (*Luther's Works*, 1958, 21.320, qtd in Clough 56).

22 Hodges, *Creatures Goodness*, 1675, 13–14. The idea that humanity's rebellion against God was mirrored in the nonhuman world's rebellion against them, while nonhuman nature still retained its essential goodness, is a commonplace in Reformed theology.

23 In his second sermon, Hodges engages in an extended discussion of hunting (29–32). He follows Luther in his qualified criticism: we should not delight in animals' violence, and hunting should not interfere with the important duties of life or pose unnecessary risks to humans. It can however produce food and clothing, and some animals seem made for it.

As Pope says (ironically) in his *Guardian* 61, who would dare "attack a Diversion which has so much Custom and Authority to support it?" (379). Hodges goes on, however, to specify the cruelties practised on domestic creatures who are part of the community, as a betrayal of responsibility and of trust.

24 Who is Himself compared to animals: "He was oppressed and afflicted, yet he did not open his mouth;/He was led like a lamb to the slaughter,/and as a sheep before her shearers is silent,/so he did not open his mouth" (Isaiah 53:7. New International Version).

References

Primary Sources

Adams, Thomas. *The Sacrifice of Thankefulnesse A Sermon Preached at Pauls Crosse, the Third of December, Being the First Aduentuall Sunday, Anno 1615. By Tho. Adams. Whereunto Are Annexed Fiue Other of His Sermons Preached in London, and Else-Where; Neuer Before Printed. . . .* London: Printed by Thomas Purfoot, for Clement Knight, 1616. Early English Books Online. www.proquest.com/docview/2240902543.

Augustine of Hippo, Saint. *Concerning the City of God Against the Pagans.* Translated by Henry Bettenson. Introduction by John O'Meara. Harmondsworth: Penguin Classics, 1984.

Bellarmino, Roberto Francesco Romolo, Saint. *Iacob's Ladder Consisting of Fifteene Degrees or Ascents to the Knowledge of God by the Consideration of His Creatures and Attributes* [De ascensione mentis in Deum per scalas rerum creatorum opusculum]. London: Printed [by Elizabeth Purslowe] for Henry Seile . . . 1638. Early English Books Online. www.proquest.com/docview/2240872481.

Calvin, John. *Sermons of Master Iohn Caluin, Upon the Booke of Iob. Translated out of French by Arthur Golding. Imprinted by Lvcas Harison and George Byshop, 1574.* Early English Books Online. www.proquest.com/docview/2264190602.

———. *The Sermons of Master Iohn Caluin, Vpon the Epistle of S. Paule Too the Ephesians. Translated Out of French into English by Arthur Golding.* London: [By Thomas Dawson] for Lucas Harison, and George Byshop, 1577. Early English Books Online. www.proquest.com/docview/2240885981.

Chaloner, Edward. *Sixe Sermons. Preached by Edward Chaloner, Doctor of Diuinitie, and Fellow of All-Sovles Colledge in Oxford.* London: Printed by W. Stansby, 1623. Early English Books Online. www.proquest.com/docview/2240868389/.

Draxe, Thomas. *The Earnest of Ovr Inheritance: Together with a Description of the New Heaven and of the New earth, and a Demonstration of the Glorious Resurrection of the Bodie in the Same Substance. Preached at Pauls Crosse the second day of August 1612.* London: Imprinted by F. K. for George Norton, 1613. Early English Books Online. www.proquest.com/docview/2240884722.

Goodman, Godfrey. *The Creatures Praysing God: Or, The Religion of Dumbe Creatures.* London: Printed by Felik Kingston, 1622. Early English Books Online. www.proquest.com/docview/2240927434/.

Hall, Joseph. *The Works of Joseph Hall, Doctor in Diuinitie, and Deane of Worcester. WORCESTER. With a Table Newly Added to the Whole Worke.* London: Printed [by John Haviland, Miles Flesher, and John Beale] for Nath. Butter [Thomas Pavier, Miles Flesher, John Haviland, George Winder, and Hanna Barret], 1625. Early English Books Online. www.proquest.com/docview/2240902706.

Hodges, Thomas. *The Creatures Goodness, As They Came Out of God's Hands, and the Good Man's Mercy to the Brute Creatures, Which God Hath Put Under His Feet. In Two*

Sermons, the First Preached before the University of Oxford, The Second at the Lecture at Brackley. By Thomas Hodges, B. D. Late Rector of Souldern in Oxfordshire. London: Printed for Tho. Parkhurst at the Bible and three Crowns in Cheapside, 1675. Early English Books Online. www.proquest.com/docview/2264187629.

King, Henry. *Two Sermons Preached at White-Hall in Lent, March 3. 1625. And Februarie 20. 1626. By Henry King, D. D. one of his Maiesties Chaplaines in Ordinarie*. London: Printed by Iohn Haviland, 1627. Early English Books Online. www.proquest.com/docview/2240860343.

Luther, Martin. *A Commentarie of M. Doctor Martin Luther Vpon the Epistle of S. Paul to the Galathians First Collected and Gathered Vvord by Vvord Out of His Preaching, and Novv Out of Latine Faithfully Translated into English for the Vnlearned. Wherein is Set Forth Most Excellently the Glorious Riches of Gods Grace* . . . [In epistolam Sancti Pauli ad Galatas commentarius]. London: By Thomas Vautroullier . . . 1575. Early English Books Online. www.proquest.com/docview/2240906390/.

———. *An Exposition of Salomons Booke Called Ecclesiastes or the Preacher. Seene and Allowed [Ecclesiastes odder Prediger Salomo]*. London: Printed by Iohn Daye . . . 1573. Early English Books Online. www.proquest.com/docview/2248578771.

———. *Luthers Wereke: Britische Gesamtausgabe, Tischreden*. 6 vols. Ed. D. Martin. Weimar: Hermann Bölhaus Nachfolger, 1912–21. No. 1150. Translated by Roland Bainton, in "Luther on Birds, Dogs and Babies: Gleanings from the 'Table Talk'." In *Luther Today*, edited by Roland H. Bainton, Warren A. Quanbeck and E. Gordon Rupp. Martin Luther Lectures, vol. I. Decorah, IA: Luther College Press, 1957, 9. Ref. in Clough, "Anxiety," 2009.

———. *Luther's Works*. Edited by Jaroslav Pelikan et al. 79 vols. St Louis, MO: Concordia Publishing, 1958. Ref. in Clough, "Anxiety," 2009.

———. *Martin Luther's Genesis Commentary (Chapters 1–4)*. Everyone's Luther. Aurora, CO: Pastor Bryan Wolfmueller, 2016. www.wolfmueller.co/everyonesluther.

———. *A Methodicall Preface Prefixed Before the Epistle of S. Paule to the Romanes Very Necessary and Profitable for the Better Vnderstandyng of it, Made by the Right Reuerend Father and Faythfull Seruant of Christ Iesus, Martin Luther; Nowe Newly Translated out of Latin into English, by W.W.* . . . London: for Thomas Woodcocke . . . 1594. Early English Books Online. www.proquest.com/docview/2240940971.

More, Henry. *An Antidote Against Atheisme, or an Appeal to the Natural Faculties of the Minde of Man, Whether There be Not a GOD*. . . . London: Printed by Roger Daniel . . . 1653. Early English Books Online. www.proquest.com/docview/2264215947.

———. *The Immortality of the Soul, So Farre Forth as It Is Demonstrable from the Knowledge of NATURE and the Light of REASON*. . . . London: Printed by I. Flesher, for William Morden Bookseller in Cambridge, 1659. Early English Books Online. www.proquest.com/docview/2240956337.

Pope, Alexander. *The Guardian*. No. 61 (Thursday, 21 May 1713). London: J. Tonson, 1714.

———. *The Poems of Alexander Pope*. Edited by John Butt. London: Methuen, 1963.

Rawlinson, John. *Mercy to a Beast a Sermon Preached at Saint Maries Spittle in London on Tuseday in Easterweeke. 1612. by Iohn Rawlinson Doctor of Divinitie*. Oxford, Printed by Ioseph Barnes, 1612. Early English Books Online. www.proquest.com/docview/2240899234.

Viret, Peter. *The Schoole of Beastes, Intituled, the Good Housholder, or the Oeconomickes. Made Dialogue-wise, by M. Peter Viret, Translated Out of French into English, by I. R.* London: Printed by Robert Waldegrave, 1585. Early English Books Online. www.proquest.com/docview/2240853102.

Webster, John. *Academiarum Examen* . . . London: Giles Cavert, 1654. Early English Books Online. www.proquest.com/docview/2240904851.

Secondary Sources

Agamben, Giorgio. *The Open: Man and Animal*. Translated by Kevin Attell. Stanford: Stanford University Press, 2004.

Bauckham, Richard. "Jesus and Animals I: What Did He Teach?" In *Animals on the Agenda: Questions About Animals for Theology and Ethics*. Edited by Andrew Linzey and Dorothy Yamamoto, 33–48. London: SCM Press, 1998.

Berkman, John. "Towards a Thomistic Theology of Animality." In *Creaturely Theology: On God, Humans and Other Animals*. Edited by Celia Deane-Drummond and David L. Clough, 21–40. London: SCM Press, 2009.

Blomfield, James Charles. *History of the Present Deanery of Bicester, Oxon: Part VII: History of Fritwell and Souldern*. London: Elliot Stock, 1893. https://archive.org/details/historypresentd00unkngoog.

Brown, Laura and Bryan Alkemeyer. "Rational Elephants or Hominoid Apes: Which Is Early Modern?" *Journal for Early Modern Cultural Studies* 13, no. 4 (Fall 2013): 61–68. www.jstor.org/stable/jearlmodcultstud.13.4.61.

Clarke, Gillian. "The Fathers and the Animals: The Rule of Reason?" In *Animals on the Agenda: Questions About Animals for Theology and Ethics*. Edited by Andrew Linzey and Dorothy Yamamoto, 67–79. London: SCM Press, 1998.

Clough, David L. "The Anxiety of the Human Animal: Martin Luther on Non-human Animals and Human Animality." In *Creaturely Theology: On God, Humans and Other Animals*, edited by Celia Deane-Drummond and David L. Clough, 41–60. London: SCM Press, 2009.

Cole, Lucinda. "Animals Studies and the Eighteenth Century: The Nature of the Beast." *Literature Compass* 16 (2019): 1–12. https://doi.org/10.1111/lic3.12536.

Davis, Ellen F. *Scripture, Culture, and Agriculture: An Agrarian Reading of the Bible*. Cambridge: Cambridge University Press, 2009. Kindle edition.

Deane-Drummond, Celia and David L. Clough, eds. *Creaturely Theology: On God, Humans and Other Animals*. London: SCM Press, 2009.

——— and Rebecca Artinian-Kaiser, eds. *Animals as Religious Subjects: Transdisciplinary Perspectives*. London: Bloomsbury, 2013.

Fudge, Erica. *Brutal Reasoning: Animals, Rationality, and Humanity in Early Modern England*. Ithaca: Cornell University Press, 2006.

———. *Perceiving Animals: Humans and Beasts in Early Modern English Culture*. London and New York: Macmillan, 2000.

———, Ruth Gilbert and Susan Wiseman, eds. *At the Borders of the Human: Beasts, Bodies, and Natural Philosophy in the Early Modern Period.* New York: St. Martins, 1999.

Harrison, Peter. *The Bible, Protestantism, and the Rise of Natural Science*. Cambridge: Cambridge University Press, 1998.

Huff, Peter A. "Calvin and the Beasts: Animals in John Calvin's Theological Discourse." *Journal of the Evangelical Theological Society* 42, no. 1 (1999): 67–75.

Huiser, Kris and Matthew Barton. "A Promise Is a Promise: God's Covenantal Relationship with Animals." *Scottish Journal of Theology* 67, no. 3 (August 2014): 340–356. https://doi.org/10.1017/S0036930614000155.

Ickert, Scott. "Luther and Animals: Subject to Adam's Fall?" In *Animals on the Agenda: Questions About Animals for Theology and Ethics*, edited by Andrew Linzey and Dorothy Yamamoto, 90–99. London: SCM Press, 1998.

Kemmerer, Lisa. *Animals and World Religions*. Oxford: Oxford University Press, 2011.

Lamb, Jonathan. "Sympathy with Animals and Salvation of the Soul." *The Eighteenth Century* 52, no. 1 (Spring 2011): 69–85. https://doi.org/10.1353/ecy.2011.0003.

Linzey, Andrew and Dorothy Yamamoto, eds. *Animals on the Agenda: Questions About Animals for Theology and Ethics*. London: SCM Press, 1998.

McLaughlin, Ryan Patrick. "Evidencing the Eschaton: Progressive-Transformative Animal Welfare in the Church Fathers." *Modern Theology* 27, no. 1 (January 2011): 121–146. https://doi.org/10.1111/j.1468-0025.2010.01658.x.

Menely, Tobias. *The Animal Claim: Sensibility and the Creaturely Voice*. Chicago: University of Chicago Press, 2015. Kindle edition.

Moore, Stephen D., ed. *Divinanimality: Animal Theory, Creaturely Theology*. Transdisciplinary Theological Colloquia. Foreword by Laurel Kearns. New York: Fordham University Press, 2014.

Morillo, John. *The Rise of Animals and the Descent of Man, 1660–1800: Towards Posthumanism in British Literature Between Descartes and Darwin*. Newark, DE: University of Delaware Press, 2017.

Moritz, Joshua M. "Martin Luther and the Medieval Saints among the Animals." *Dialog: A Journal of Theology* 51, no. 1 (March 2012): 7–12. http://doi.org/10.1111/j.1540-6385.2011.00648.x.

Nash, Richard. *Wild Enlightenment: The Borders of Human Identity in the Eighteenth Century*. Charlottesville: University of Virginia Press, 2003.

Patton, Kimberley C. "'He Who Sits in the Heavens Laughs': Recovering Animal Theology in the Abrahamic Traditions." *Harvard Theological Review* 93, no. 4 (October 2000): 401–434. www.jstor.org/stable/1510165.

Preece, Rod and David Fraser. "The Status of Animals in Biblical and Christian Thought: A Study in Colliding Values." *Society & Animals* 8, no. 3 (2000): 245–263. https://doi.org/10.1163/156853000511113.

Quinsey, Katherine M. *Animals and Humans: Sensibility and Representation, 1650–1820*. Oxford: Oxford University Studies in the Enlightenment, 2017.

———. "Little Lives in Air: Animal Sentience and Sensibility in Pope." In *Animals and Humans: Sensibility and Representation, 1650–1820*, edited by Katherine M. Quinsey, 141–172. Oxford: Oxford University Studies in the Enlightenment, 2017.

Raber, Karen. *Animal Bodies, Renaissance Culture*. Philadelphia: University of Pennsylvania Press, 2013.

Schreiner, Susan E. *The Theater of His Glory: Nature and the Natural Order in the Thought of John Calvin*. Studies in Historical Theology 3. Durham: Labyrinth Press, 1991.

Shannon, Laurie. *The Accommodated Animal: Cosmopolity in Shakespearean Locales*. Chicago: University of Chicago Press, 2013.

Shevelow, Kathryn. *For the Love of Animals: The Rise of the Animal Protection Movement*. New York: Henry Holt, 2008.

Simons, John. *Animal Rights and the Politics of Literary Representation*. London: Palgrave, 2002.

Steiner, Gary. *Anthropocentrism and Its Discontents: The Moral Status of Animals in the History of Western Philosophy*. Pittsburgh: University of Pittsburgh Press, 2005. Kindle edition.

Thomas, Keith. *Man and the Natural World: Changing Attitudes in England 1500–1800*. Oxford: Oxford University Press; London: Penguin Books, 1983. Kindle edition.

Watson, Robert N. "Protestant Animals: Puritan Sects and English Animal-Protection Sentiment, 1550–1650." *English Literary History (ELH)* 81, no. 4 (December 2014): 1111–1148. www.jstor.org/stable/24477772.

Wolfe, Cary. *What Is Posthumanism?* Minneapolis: University of Minnesota Press, 2010.

Wolloch, Nathaniel. "Animals in Enlightenment Historiography." *Huntington Library Quarterly* 75, no. 1 (2012): 53–68. https://doi.org/10.1525/hlq.2012.75.1.53.

8 Bandits

John Van Rys

The video footage was dark and grainy, but the shape exiting the coop door was unmistakable.[1]

"Egg stealer! Chicken killer!"

Evan watched the video over and over on his laptop screen at the kitchen table as Arthur played it in a short loop. Arthur, just finished grade 9 that summer, had set up a small camera in the coop after Evan had complained of broken eggs and missing chickens, of missing eggs and broken chickens. For several days he'd gone out to the coop to find egg production way down, brown shells broken and hollowed out, feeders knocked over and raided. Worse, he'd found serious disturbances in the sand, feathers and down scattered about, plus the remains of a chicken wing in the outdoor run—clear evidence the hens in his flock were being picked off.

He asked Arthur to freeze the image on the dark shape exiting the small door he kept open during the day. "Little bastard!"

"It's probably a female, Dad."

"Bitch! You're going down."

"Calm your shit, Dad. It's just a raccoon."

"Watch your language." He looked at Arthur. "There's no such thing as *just* a raccoon. And what do you mean, calm my shit? My shit is calm, very calm." He turned his attention back to the screen. "Run it again." The video looped, the raccoon escaping again and again and again. "Bloody thieving bandit." He felt his twin, Evil Evan, expanding in his chest, blocking out Even Evan, his mild-mannered self.

He loved his hens, maybe a bit too much according to his wife Mae. "You spend more time with them than you do with me," she'd say. "You might as well have a mistress." When he'd get a new batch of chicks—yellow, brown, or black puff balls with beaks and black-bead eyes, all perched on tiny legs and claws so large for their bodies—she'd complain he fretted more about them than he ever did about his own kids growing up, all four of them combined. "The chicks are less trouble," he'd reply. "They don't turn into teenagers." He got them day-old from the farm store in nearby Windermere and tended them carefully in the brooder under a heat lamp he constantly monitored until their feathers came in. That's when they were old enough to go out to the coop he'd built in the barn, where he'd keep them separate from the adult hens until the young birds were old enough to fend for themselves.

DOI: 10.4324/9781003366744-8

His adult ladies were prone to pecking the little ones, even killing them—an old, old story about the female of the species, he thought.

He loved to feed the hens and fill their waterers, they seemed so eager to eat and drink and make merry and lay eggs. He'd walk carefully through the coop so as not to step on any of them, for they were skittish and flighty—bird-brained, after all—and were likely to get underfoot. When he brought winter snow into the coop on his barn boots, they'd gather around and peck at it. If he tried to gather eggs from beneath them, they were likely to peck his hand. He'd go back up to the house and complain to Mae. "Look at me. I'm hen-pecked."

"It must not be working very well, judging from my honey-do list."

"I'll get to it after we finish everything on our honeymoon list." He liked to reference this list occasionally, constructed out the sexual adventures related by Larry and Caroline Laurence at the Talisman, as well as a copy of *The Kama Sutra* he and Mae borrowed from the bridal suite. They hadn't returned it yet.

"In your dreams," Mae replied.

He'd share this hen-pecked husband joke with his kids and the rest of his family. They complained it was a groaner. He'd try it out on his egg customers. "Look at these scars," he'd say. "You can tell I bled for these eggs."

None of his hens were white—no boring Leghorns for him, no white eggs. Over the years, he rotated between getting Rhode Island Reds and Plymouth Rocks, along with breeds whose names seemed strange, like Black Sex Link. Such a hot name for a chicken, I have to get that breed, he'd thought.

And the brown eggs they laid were beautiful, not like store-bought eggs at all. No, his weren't blandly uniform in colour and size, robotically manufactured at some factory farm. His hens laid eggs of different shades and sizes, some with odd shapes and wobbly ridges and spots, some pure brown and others dappled, occasionally one with a double yolk. They were all lovely variations of ovoid—he'd looked up the word for egg-shaped, from ovum for egg. He corrected people when they said *oval*. That's 2-d, he'd say. *Ovoid* is the real deal, 3-d. In his personal cosmology, the universe was ovoid. God had laid a giant egg and was brooding over it for billions of years. Any day now it might hatch into something heavenly. His hens' eggs were tiny replicas of this universe. They came out protected by a liquid sheen, a barrier against bacteria and antimatter. After the egg dried, he would just brush off any sand stuck to it. Maximum protection. No chinks in the egg's ovoid armour. Fragile armour, of course, easily cracked. But armour nonetheless, he chose to believe.

Occasionally when he scratched his ears or rubbed his eyes or ran his hand over the little bit of hair he had left, it occurred to him the human head was a bumpy ovoid container for the brain and its various attachments. Here was another mysterious connection between himself, the cosmos, and eggs.

His eggs did taste like heaven. It was true. Several people had told him as much. He had a sign at the road for neighbours and cottagers: "Free Run Omega eggs. $4 a dozen." He brought eggs to coworkers at Wheel and Barrow Chartered Accountants. (He thought of them as his egg clients, not customers; he himself was their egg broker.) And he gave gifts of eggs to family when they came to visit. "There's no comparison," everybody said. "Your eggs are absolutely delicious."

He'd thank them and say "Happy hens mean happy eggs." He did sometimes forget he wasn't the one laying them. It felt good to take credit—but a little weird.

He, Mae, and the kids had made a family project of it when he'd started talking about getting hens. Mae bought a couple of how-to books on chickens, which they studied together as a form of marriage counselling—growing closer by learning how to make a home for hens, rekindling the fire by raising fowl. Together with the kids, they framed out a coop inside their old barn and covered it with sturdy wire called hardware cloth. They covered every possible opening to make it an impenetrable chicken fortress, a Fort Knox to protect all those golden yolks. They'd installed a small door in the barn wall so the hens could go out during the day, built nesting boxes out of plywood and mounted them on the wall, and fashioned roosts out of an old wooden ladder. Mae read on the Internet—she'd found some interesting chicken chat rooms—that sand was the perfect litter and bedding for chickens. But it couldn't be sandbox sand, it had to be construction-grade sand. So they ordered a dump truck load, and filled wheelbarrow after wheelbarrow to cover the floor of the coop and bed the nesting boxes.

He'd learned the hard way the coop needed to be a fortress. The summer they started their chicken project, his daughter Lizzie was doing volunteer hours at a vet clinic in London. She'd just finished Grade 9. One day she brought home three young chickens from the clinic, abandoned birds dropped at the door in the early morning hours, probably by some urban farmer who thought he could raise chickens in an apartment. They decided to temporarily keep the chickens outside in an old wire dog crate until they could build a coop. But that night in bed Evan heard a commotion outside. By the time he'd pulled on some sweat pants and a T-shirt, grabbed a flashlight, and gone to investigate, two of the young chickens had vanished and the third was lying dead on the lawn outside the crate—all of them having mysteriously escaped the narrow bars, a sinister magic trick. He swung his flashlight beam in an arc across the lawn and caught the bandit waddling away with its plunder. In the light, it paused and looked back, its masked eyes aglow, then continued its escape.

While they were waiting for their first order of day-old chicks to arrive, they built their own brooder—a plywood box on wheels with a two-by-four framed lid covered in wire so they could watch the chicks but keep them safe from their dogs and cats, and hinged so they could lift it easily to feed and water the little peepers or catch and cuddle them. The inside bottom was protected by linoleum peel-and-stick tiles, a blue design left over from renovating the laundry room, plus several inches of pine shavings. They rigged up an arm from which to hang the heat lamp, the height adjustable so the warmth would be Goldilocks right as the birds grew.

That was about a decade ago, so the brooder had helped raise dozens of chicks into hens. To Evan each bird was vitally important, though he wasn't foolish enough to name them. They weren't just his hobby, not even just his passion. No, his free-run chickens were saving the planet, one egg at a time. He was doing his bit for animal rights, too—no cages for his hens.

He'd thought about installing cameras inside the coop and overlooking the run so he could live-stream his hens on the Internet. He'd considered writing a chicken blog, opening Instagram and Snapchat accounts, and setting up his own YouTube channel, becoming what the kids these days called a tuber. Who needed cat videos when you had chickens—hours and hours of entertainment. Could there be anything more comical than the way they ran to him when he brought them the kitchen scraps for a treat, racing like little feathered raptors, flapping their ineffectual wings. He'd open a Twitter account and send out regular tweets, real tweets from his chicks—hah! He'd be rich and famous, and so would his birds.

Never mind that he wasn't making any money selling eggs. That wasn't the point. He could figure it out if he wanted to. After all, he was an accountant. He knew he should probably charge more than he did just to break even, but he didn't care. He wasn't interested in a cost-benefit analysis. He sometimes said the egg money was his retirement fund, but that wasn't true. It disappeared into the daily errands of living. He loved the fresh eggs himself, that was true. But that wasn't quite it either. He couldn't really explain it. Maybe it was like adoption. His two youngest kids, Arthur and Max, were adopted. He and Mae had adopted rescue horses and dogs, too. Maybe he just felt this urge to adopt chicks, to be a chicken dad. Did that make him a rooster, a strutting Chanticleer? Or could it be that because he was a guy, it was a case of egg envy?

Maybe it was just the birds themselves, living creatures. The mystery of their growing up, the lives they led, their habits and behaviour. The satisfaction of watching them flourish, the worry they wouldn't. Or it could just be about feeding them and seeing what magically popped out the other end, accompanied by some significant clucking.

But now the magic was threatened. A raccoon had discovered a weakness in Fort Knox, the little door to the run, and was breaching its defenses to plunder and pillage at will. The creature was trespassing. He'd been taught to pray, "Forgive us our trespasses as we forgive those who trespass against us," but he wasn't having any of that. His property was being trespassed; his world felt violated.

He shut his laptop and without a word to Arthur headed out to the barn. With a corn broom he shepherded the hens out of the run and into the coop. He closed the run door, then went inside and latched it with the sliding bolt. "Sorry, ladies," he said. "You're in witness protection for now."

Despite his initial anger and urgency, he took more than a week to get around to buying a live trap. His foe was cunning, and he didn't relish taking it on in battle. Instead, he looked for the easy way out. Keeping the hens in the coop permanently would solve the problem, but he couldn't stand the idea of taking away their freedom. They'd literally be cooped up if he did that. He considered hiring an exterminator or wildlife removal service, but he and Mae had moved to the country to be closer to nature. This solution seemed a copout, a city-slicker's solution. Besides, being Dutch-Canadian he was too cheap to lay out dough on getting rid of a pest. He seriously weighed poison or a shotgun as options—at least for a moment—but in the end rejected them. He'd studied just-war theory in university, but at heart he was a pacifist—though he did have to admit his principles were being sorely tried. He thought through these options, and discarded each in turn.

Each day, Mae asked him if he'd picked up a trap on his way home from work. She'd done the research for him, found one at the farm store in Windermere. Finally, on the Friday he did the deed and laid out the cash. It was almost two weeks since the video had caught the bandit on tape.

That evening he hardboiled a half dozen eggs. He didn't want to, but he knew he had to sacrifice some yolks for the greater good. He set up the trap inside the barn, right beside the coop. Inside the far end of the trap—a rectangular wire box with an opening on the near end—he placed two hard-boiled eggs in a paper bowl, then pulled back the handle on the top that set the trap. When the bandit walked into the trap, it would step on the metal trigger and release a flap behind it. Ingenious and humane, he thought, as he stood and examined the trap. Work your magic tonight.

The next morning felt like Christmas. He could hardly wait to get up and see what present the trap had left him. He was awake long before Mae and the kids, pulled on his barn clothes and his boots, and strode out to the barn while leaving tracks in the dew-soaked grass. He quieted himself as he approached the barn door, then peered in. In the early morning light the barn was dim, but he could see a dark shape in the sprung trap. An unmistakable shape.

He flipped on the lights and calmly approached the trap. He saw that both eggs were gone. The raccoon was sitting with its rear pressed against the trap's flap. Unblinking, it stared at him. He could swear the look on its face was one of patient resignation. It looked somewhere between calm and dejected, almost embarrassed having been caught wearing its bandit mask and prison-striped tail. But when he came closer it began pulling at the bars with its clever paws, its thin black fingers. He stopped, afraid. The phrase "primitive and irrational" popped into his head, but the fear remained.

He returned to the house, shook Mae awake, and explained they had a trapped raccoon to deal with. "You mean *you*, right?" was Mae's reply. "You realize I was having a lovely dream about horses. I was riding a sorrel mare on a Sable Island beach. I'd gotten special permission from the Prime Minister, who wasn't wearing a shirt. Now I'll never know how the dream ends."

"Sorry, but there's a nightmare out in the barn, and I could use some help figuring out how to get rid of it."

Mae emerged from under the sheets and followed him downstairs. While he made them some tea, she settled herself in the living room, opened her laptop, and started researching what to do when you have a live raccoon in a trap.

When he brought in the tea, she was ready. "It says here you need to relocate the raccoon at least five miles away, though ten is better—else it'll find its way back home."

"Clever critter. Guess I'll have to drive around and find a secluded spot. It wouldn't be neighbourly just to drop it off at someone else's house or take it to the Timmy's in town."

As he sipped his tea he considered the possibilities. Most of the land around them was settled into farms—cash crop and dairy, some beef cattle, further east some vegetable fields and orchards, ginseng in what used to be tobacco country. Maybe he could release the raccoon in one of the few remaining tobacco fields so it could wreak havoc there. Perhaps he should take it to London, release it there under the cover of darkness. Lots of trees in the forest city, not to mention garbage cans and dumpsters. Wasn't the city their natural habitat now, not the country?

Better yet, why not take it to Toronto? They love raccoons there. The thought sent him on a wild, fantastic road trip. He sees himself rolling down the 401 in the fast lane, the raccoon in the back safely belted into an old car seat from the kids, peering out the window, enjoying the changing view from farmland to big city. He takes the critter downtown, checks it into the Royal York. The next day, it does all the tourist stops. At the ROM it learns natural history. At the AGO it enjoys Group of Seven landscapes. It takes in a ball game, rides the subway, visits some friends at the zoo, checks out some fish at Ripley's, and scales the CN Tower. City folk video the whole climb, forgetting to turn their phones sideways, and post it on the Web. The whole world watches as the little bandit makes it to the Edgewalk deck and is invited into the restaurant for lunch. Later it checks out of the hotel without paying the bill and hooks up with some long-lost cousins. Together, they tear up the town—or at least several garbage cans with raccoon-proof lids. They're the new kings of Kensington.

Not a bad idea, Evan thinks, to take it to the city. After all, the raccoon's dominion was now everywhere—east, west, north, and south, from sea to shining sea. How far north, he wasn't sure. Probably not to the Arctic Ocean, so not sea to sea to sea—but pretty close. Maybe he should start a campaign to make it Canada's national animal, replace the humble beaver. That might raise our profile around the globe, show a different side of the national character.

He imagines taking the raccoon to the beach instead—Port Stanley might be good, Port Dover even better. He sees himself on a Harley, the helmeted raccoon behind him holding onto his waist with its clever hands. They're on their way to a Friday the Thirteenth party. It could hang out with fellow banditos, clean those hands in Lake Erie, dine on smelt or Lake Perch, or sun itself on the beach while looking for a mate. He sees himself heading southwest from there and riding with it to Point Pelee, making it the southernmost raccoon in Canada. He could introduce it to Margaret Atwood, who'll invite it to have tea and biscuits. They'll share tips about bird watching and nesting sites before it swims to Pelee Island.

He comes back to his senses as he takes his final sip of tea. "I think I'll take it to Backus Woods," he says to Mae. "That's more than forty miles away. Plus, it's secluded and it's a conservation area. It's exactly where a raccoon should live—its natural habitat if it hadn't been for people."

"Just one more story of colonization."

"Colonization? You think we're the problem, not the raccoon?"

"I wouldn't put it like that. But you might look at it from the raccoon's point of view."

The raccoon's point of view? Did it have one? "I suppose you mean live and let live. Coexist peacefully. But if we did that, we'd be overrun by them. The barn would be filled to the rafters and there wouldn't be one chicken or egg left—in the coop or on the planet. Life's most profound philosophical question—which came first, the chicken or the egg?—could no longer be asked, let alone answered. They'd colonize us, not the other way around. The raccoons will take over the world if we let them."

"I don't think you really have to worry about that, judging by the number of them that get hit by cars."

"Fair enough, but if we don't set up boundaries and keep them in their natural habitat, pretty soon we'll be just like Toronto. Before you know it, raccoons will outnumber people. They'll infiltrate city hall. I wouldn't be surprised if one day a raccoon was in the mayor's chair. Of course, it might do a better job of fixing Toronto's housing problems. But they'll stop at nothing. Queen's Park, the Parliament Buildings. Just look at politics south of the border. It hasn't gone to the dogs; it's gone to the raccoons."

"You're getting just the slightest bit carried away, don't you think?"

"Not a bit. You know what they say, 'Good fences make good neighbours.' There's wisdom in that thought when it comes to raccoons."

"If I thought you were being serious, I'd have to have you committed."

"You haven't looked into a raccoon's eyes. It's all there—thwarted ambition, criminal intent." He got up to go. "I should head out before the roads get busy. I'm not sure I want witnesses to this raccoon relocation."

"Good luck. If you're not back by noon, I'll send out a search party."

On his way outside he grabbed his barn gloves and coat for protection, then found a tarp and a pitchfork. He'd decided to use Mae's hatchback for the relocation. He told himself it was the only practical vehicle for the job, as the trap might not fit in his trunk and would definitely get his back seat smelly and dirty. Plus their old F-150 was low on fuel and reliability at the moment. He put the tarp down in the back of her car and carefully laid the pitchfork over the seats.

In the barn the raccoon appeared not to have moved. It stared at him as he walked over, adjusting its position only slightly to face him squarely. He picked up the trap by the handle and began walking as quickly as possible to the car, holding the trap away from his leg, fearing those hands that could magically pull things through bars. The raccoon began lunging at the trap's back end, trying to break out. The force and weight knocked Evan off balance. He tripped and dropped the trap.

The door held. Thank you, Jesus, he thought. He found himself on the ground, eye level with the raccoon. For a startled moment, he met its dark eyes ringed in black fur. There, he saw an inscrutable life, as he had years before in the eyes of a rescued Percheron named Finnegan. Evan noted the brown-black stripe running between the raccoon's eyes from its nose to its forehead and was surprised by the white of its whiskered snout, more white around the bandit black, a rim of tan-white on the edges of its ears. The white gradually blended into the brown and black fur on its body, heavy all over until it smoothed out just above its hands and feet. The hands themselves were black, the fingers distinct—thin and clawed. The moment passed; the raccoon backed away from his face, hissing. Guess it didn't like what it saw. What did it see? He considered his balding head and big nose, his beard, his teeth, his eyes behind the barrier of his black-rimmed glasses. To the raccoon, a bandit's face?

He got up, brushed himself off, and looked down at the creature in the cage. He picked up the trap and now walked slowly and carefully to Mae's car. He gently placed the trap on the tarp and closed the hatchback without slamming it.

The roads to Backus Woods were quiet, but the trip was anything but restful for Evan. Before they'd even gotten down the long laneway to the road, he could hear scrabbling from the back of the car. He had a vision of the raccoon escaping the trap, its head popping above the back seat. The creature climbing over and jumping on top of his head, a living coonskin cap pulling off his glasses and nibbling on his ears. The rational part of his brain kept saying the trap door would hold, it was properly designed and manufactured. The irrational part wasn't convinced.

When he got to Backus Woods he looked for a secluded spot, a secondary road or trail that would give him enough time to release the raccoon without the interference of passersby or park rangers. For all he knew, what he was doing was trespassing since he wasn't using the park entrance. He pulled off at a spot where the woods seemed to have some dense undergrowth and put on his hazard lights. He got out, donned his barn coat and gloves, and opened the hatchback.

For the whole trip the raccoon had been scrabbling about in the trap, pulling in and shredding pieces of the tarp. Mae's car was a mess. Evan lifted the trap out and began yanking the tarp back through the bars. When he'd extricated the tarp from the trap, he placed the trap off the road's shoulder at the edge of the ditch, the door facing the woods. He retrieved the pitchfork from the car. Armed with the pitchfork in his left hand, his heart thumping in his chest, he placed his gloved right hand on the latch that opened the trap door. He pulled hard and the flap snapped up. The raccoon shot out, raced across the ditch, and disappeared into the undergrowth—all in a couple of seconds. Startled, Evan jumped back into the roadway. The last he saw of the raccoon were the shrubs rustling as it made its way into the forest.

Enjoy your new home, he thought. Y'all don't come back now, y'hear?

He looked around at the deserted road and the quiet woods on either side. Except for morning bird song, all was silence as he stood in the middle of the road, pitchfork in hand. No breeze stirred the leaves and needles. He felt watched. Quickly he placed the trap and the pitchfork back in the car, slammed the hatchback down, slid into the driver's seat, and pulled back onto the road, sweating in his barn coat and gloves. It was several minutes before he realized his hazards were still blinking.

That evening he reset the trap with more hardboiled eggs. Better make sure there aren't any more raccoons, he thought. Are they solitary or sociable creatures? He wasn't sure. When he returned from the barn, he called his son Alex and turned the day's events into a comical tale with himself as the mock-hero. "Picture me," he said, "armed with my pitchfork at the edge of the wild woods, the monster about to be freed at my feet. I'll tell you, that raccoon was a lot scarier than the killer rabbit in *Monty Python*. And there I was with no holy hand grenade in sight."

Despite such lighthearted tale-telling, that night Evan had nightmares. He was in the barn. It was murky dark with just thin ribbons of moonlight coming through the windows and doors. The hens were unsettled on their roosts. In his nightmare they too were having nightmares. There were nightmares within nightmares. The trap was sprung with a raccoon in it. As soon as he ran it to Backus Woods he'd come back and the trap had another raccoon in it, even though the trap was still in his hand. As fast as he removed raccoons, the trap filled up—sometimes with four or five squeezed into it at the same time. He ran out of room in the woods—the trees were full of them, the forest floor was crawling with them—so he started putting them in the basement and the attic. In the nightmare, Mae told him he'd have to move out, their marriage was done, finito. There wasn't room for both him and the raccoons, and they were better company. She'd always liked bad boys, banditos.

He woke up in the dim light and looked at his sleeping wife. What was it about women and bad boys? Is that really what they wanted? Where had goodness gotten him, all these years? Of course, he'd gotten her pregnant when they were dating—going steady it was called then—and they'd eloped, running away to the Talisman Ski Resort in Beaver Valley. Didn't that make him a rebel? But that was pretty much it for his bad-boy behaviour, and it was so long ago now—more than thirty years. He'd turned into a mild-mannered accountant, the Clark Kent of the spreadsheet. Maybe for the sake of his marriage he needed to revive his inner bandito. Should he buy a motorcycle, grow a ponytail with the wisps of hair on the back of his head, get one of his nipples pierced?

It was way too early for such questions. He was much too groggy to sort out the puzzle of his marriage, the tangled love knot he felt for Mae, a knot that went all the way back to a bike ride, their first date. Not to mention it was a knot threaded through with their four kids, with decades of being called *Dad*, that word still a mystery to him.

After this restless night, he was exhausted. He was afraid to go out to the barn. He needed to give himself a vigorous face wash before he could investigate what was out there on this quiet Sunday morning. When he put on his barn boots, his feet felt heavy; they swung like iron pendulums as he trudged to the barn.

He peered into the gloom before turning on the lights. The trap was once again sprung. He could see shapes moving within it. But what was that he heard? It sounded like the chirping of birds. When he flipped on the lights he was startled to see two baby raccoons huddled together in the trap. Above them he saw another baby crawling up the outside of the coop, the chickens on the roosts eyeing it curiously. One or two started pecking at the hardware cloth. The baby fell to the floor and starting climbing again. Hearing chirping to his right, he turned to see two more babies climbing up a post from one of the horse stalls, seeking higher ground away from him, a threat. All five were now chirping their alarm, calling their mother. It suddenly occurred to him, their mother was in Backus Woods. *It* was a *she*. He'd been thinking *it* all along, and here he was confronted with children, her babies.

He had to act quickly, before the three loose babies got up into the loft. He grabbed the metal dog kennel he'd used many years ago for those first doomed chickens and filled the bottom with pine shavings. Using a step ladder, he first climbed up to the two strays above the horse stalls. When he tried to grab one by the scruff of the neck, it turned and nipped his barn glove. He changed tactics, picking up the baby by its tail and holding it aloft until he could stow it in the dog crate. He repeated this procedure with the other loose babies before retrieving the two from the live trap. Now all five were chirping, exploring the crate's bars, and crawling over each other.

For the second morning in a row, he roused Mae from her dreams. What was to be done? He made such a commotion that Max and Arthur emerged from their bedrooms as well. When they got the story, they headed out to the barn to see for themselves.

"Leave them in the crate for now, until we decide what to do," Evan shouted after them.

Mae went on the Web, looking for answers. While she did that, he called Lizzie. By this time, she had started vet school after studying animal biology at Guelph and working summers at a nearby vet clinic.

"I was afraid of that," she replied when he'd told her about the appearance of the babies. "The babies are called kits, by the way. I figured you'd caught a female that had made a nest somewhere in the barn."

"So now I'm a home wrecker?"

"Sorry, Dad, it's the circle of life."

Mae interrupted. "Let me talk to Lizzie. I've got some information here I'd like to check with her." He passed Mae the phone and went into the kitchen, putting on the kettle and listening to one side of the conversation between his wife and daughter. He didn't want to participate in the circle of life. He wanted life to be ovoid, complete and contained with a rich yoke at the middle.

By the time Mae got off the phone, the tea was ready. "All right. I've got some numbers here we can call. There's the SPCA and a couple of wildlife rescue centres in the area. It also says here kits that size can eat some dog kibble. Lizzie said to soften it and mix it with some scrambled eggs. Give them a dish of water too. While you're doing that, I'll make some calls."

He scrambled up a dozen eggs while the kibble was softening in a dog dish. When he'd mixed it all together, it looked like brown and yellow vomit. He took the dish of food, a pitcher of water, and an empty dog dish out to the barn.

Max and Arthur had taken kits out of the crate and were petting them. "What are you doing? I said to leave the kits in the crate."

Arthur answered, "What do you mean, kits? Aren't they pups or cubs or coonlets?"

"According to your sister, they're kits. You still haven't answered my question."

It was Max's turn. "They wanted to play, Dad. I mean look at how cute they are. They were begging to come out."

"I don't care if they said 'pretty please,' you shouldn't have taken them out. What if they're carrying diseases? Put them back in the crate. I've got food and water for them."

The boys popped their playmates back in, and held the door open for him. He first put the food in, followed by the water. Even before he'd gotten the door closed, the kits crowded around the dish of water and began dipping in and washing their hands.

"Look at that," Max said.

"Cool," Arthur added. "Why are they washing their hands?"

"They're known for doing that, but I'm not quite sure what it's for. Instinct, I guess, or maybe they've just got good manners, washing their hands before they eat. I know a couple of boys who could take some lessons from them."

"What do you mean? I wash my hands every time I go to the bathroom and before every meal. It's part of my religion." Evan knew Arthur's claim to be a whopper but said nothing. Max remained silent on the subject. As Evan and the boys watched the kits washing and drinking, he thought about the life contained within their small bodies. The bandit masks were already there, the ringed tails. Such tiny creatures, yet trespassers in training. How did they fit into the bigger scheme of things? Then again, how did he?

When he and the boys got back to the house, Max related what they'd seen the kits do. "Can we keep them, Mom? You know I've wanted my own dog for years, and you've always said no. But you didn't say anything about a pet raccoon."

"Are you nuts? What are they going to eat, your dad's eggs?"

"What's wrong with dog food and scrambled eggs? I'd make it for them."

"You do realize that dogs and raccoons are enemies?"

"I'd train them to get along."

"Kinda like you and Arthur get along?"

"That's different. We're brothers. Brothers are supposed to fight."

"The answer's still no. Raccoons don't make good pets. That's what the Internet says, and your sister agrees. They're hard to raise and they pretty much stay wild. Not to mention that they're ingenious with their hands. You'd never be able to keep them out of the kitchen cupboards. Every time I turned around, they'd be in the fridge or the garbage can."

Evan had to agree with Mae. The house had already gone to the dogs and cats, and sometimes to the chicks when they were in the brooder. It would be disastrous to bring raccoons into the mix. "Besides," he added, "you boys will be off to college in a few years. Your mother and I would be stuck with them."

Max turned to Arthur. "It was worth a try." The boys headed back outside to check on the kits.

"Any luck with your phone calls?"

"Not yet. That's no surprise though on a Sunday morning."

When he went to church he couldn't keep his focus on the service. His mind kept wandering to the dog crate full of kits in the barn. He could hear their noisy chirping, the racket of little furry birds. They're what he saw when the little kids went forward for children's time and headed off to Sunday School. He prayed to God, the creator of this beautiful ovoid universe, for wisdom. He prayed that the God of children and sparrows and lilies of the field would also be the God of chicks and kits.

That afternoon, Mae got through first to the SPCA and then the wildlife refuges. The news wasn't good. The SPCA didn't take in raccoons, and the ladies at the refuge centres both said they were full. Worse yet, they said the rescue centres all over Ontario were full of raccoons. There was no room at the inn, an old, old story. They did offer the best advice they could under the circumstances. Evan would have to take the kits to the spot he dropped off the mother and hope for the best.

So he scrambled a dozen more eggs, depleting his supply almost to zero, and filled a bucket with dog kibble. He loaded the crate full of kits into Mae's hatchback and retraced his route through the lazy Sunday afternoon countryside, the summer fields of wheat and corn and soybeans, the woodlots, the orchards and vegetable patches, the little towns and crossroads sleeping in the heat. All the way to Backus Woods the kits kept up their chirping.

When he arrived at the same spot where he'd dropped off their mother the day before, he pulled over and put on his hazards again. He grabbed the bucket of food and waded through the ditch grass and the curtain of undergrowth. On the other side he found the woods, with open forest floor, spongy with humus and green growth. He looked around nervously, certain he was now trespassing the boundaries of the park. He tossed the bucket of food onto the ground and returned to the car. Without gloves or a pitchfork, he pulled out the crate full of scrambling, noisy kits. He carried it awkwardly down into the ditch and through the undergrowth, snagging his clothes and the crate, scratching himself on brambly branches. When he broke through to the open woods, he put the crate down by the food and opened the door. The kits clung to the inside bars. He tried shaking the crate upside down to dislodge them. In the end he had to reach in and pull them out. But once out they all headed for trees and starting climbing to the safety of branches.

I guess that's a good sign, he thought, as he latched the gate and carried the crate back to the car. He pulled back onto the road and the chirping faded away.

When he got home he called Lizzie and described the kits' climbing into the trees. "That's good, Dad. It means they're old enough to find safety. Hopefully, their mother will hear them. Either way, there's nothing you could do. Either their mother will find them, or a predator will. It's not pretty, but that's the nature of things. It's the circle of life."

"You mean the circle of death."

"Remember the mother was stealing your eggs and killing your chickens."

"To feed her own young, though."

"True, but think of it from the prey's perspective, not just the predator's."

"Hmm . . . your mom suggested I look at it from the raccoon's point of view."

"She's right, but so am I. You're talking about a whole ecosystem, the biome our little farm's part of, us and our species included. Look at it from both perspectives and throw in your own—homo sapiens, you know, wise person."

"Wise person? Does the name really qualify for our species? Personally, I'm not sure I'm smart enough for that. It might be more than I can manage. I just want the world to be simple and whole, a perfect ovoid."

"Ovoid?"

"Yeah, egg-shaped." And he launched into a description of his beautiful cosmology until Lizzie was laughing. "Ovum, ovoid, oval, ovary—they're all part of one beautiful eggy world."

"You're such an egg-head, Dad."

"Especially now that I'm bald."

"Problem is, you've forgotten about the pesky sperm."

"True. The intrepid little swimmer who crosses oceans just to break an egg. He's a necessary evil, I guess. Come to think of it, he's the one who started it all between your mom and me. The little troublemaker got her pregnant, which led to eloping, which led to your oldest brother."

"Nothing but trouble."

"Don't forget a little swimmer led to you too."

"Like I said, nothing but trouble."

"Then there's your younger brothers."

`"Double trouble."

"All because of that little swimmer. Guess we can blame everything on him."

That evening, Evan reluctantly reset the trap. He'd come to see that using a live trap could lead to unwanted encounters, complications. But if there were more raccoons he needed to face them.

The next morning before heading to work he was relieved to find the trap empty, but he kept the hens inside, just in case. At work, he told a white lie, said he had an appointment that afternoon so he could leave work early. He did think of it as an appointment as he drove out to Backus Woods. He felt foolish, didn't want Mae to know about it because he wasn't sure how to explain what he was doing. He was still haunted by the chirping of the kits.

He found the spot, pulled onto the shoulder, and put on his hazards. They were necessary now, given the traffic flowing in both directions. He realized he wasn't dressed properly for trespassing the park's boundaries again, in his dress shoes and dress pants, his shirt and tie. But he needed to see. He paused until the traffic let up, sheepish under the curious stares of passing motorists. He descended into the ditch, slipping in his shoes and landing on his elbows, staining his sleeves green. He crouched and worked his way through the undergrowth, stumbling onto his knees and muddying his trousers.

When he emerged beyond the curtain into the open wood, all was quiet. He looked up into the green canopy but saw no movement. He heard no chirping of furry birds. Where he had thrown the bucket of food was now ordinary forest floor. He wondered for a second if he was in the wrong spot, but no. This was the place. The food had all been eaten. Nature in some form or other had consumed it. He hoped Nature had taken care of the kits, too. But there was no knowing what had happened to them. The woods remained silent on their fate.

At that moment a breeze stirred up the canopy of leaves and found his sweating face. A dozen or more birds erupted noisily from the branches above him, where until that moment they had been silent and still. He wondered at their fierce and fragile life within the shelter of the anchored, ancient trees. He knew little about the types of trees he was looking at, even less about the birds. Perhaps he could still learn. It wasn't too late.

At home, he put his dirty clothes in the laundry. Days later, Mae would notice and ask him how he had soiled his elbows and knees. He would do his best to explain why he had to go back. "Okay," she said, "but you had me worried. I thought you'd had some kind of outdoor rendezvous. You know, *au naturel* with some wild woman, one of those single-breasted amazon warriors you're always talking about."

"It was a rendezvous, I guess, just not that kind." He tried to laugh off her worry and his own behaviour.

In the days before Mae's discovery, he kept the hens in the coop and the trap set. Each morning, the trap was empty. On the Saturday morning—a week after taking the mother raccoon to Backus Woods—he opened the coop door and released the hens into the run. He treated them to piles of fresh grass clippings and kitchen scraps he'd saved up, along with a couple of cabbages he'd bought. For a good while he watched the frenzy of pecking. It was good to feel hen-pecked again.

He went to the library in town and signed out a book on birds and another on trees. Next he picked up a poultry catalogue from the farm store. At home he spent time paging through the books, looking for the birds and trees from Backus Woods, others from his own farm. He spent some time paging through the catalogue, studying the equipment he could buy and the breeds he could raise. Later, he'd talk with Mae and they'd fill out the order form together. After all, he had to bring his flock back to its optimum size. Eggs were needed. People were depending on him. And so was the ovoid universe.

Note

1 This story first appeared in *Moonshine Promises* by John Van Rys, published by Wipf and Stock (Eugene, OR) in 2021. It is reprinted by permission of the author and the publisher.

9 Paragon of Animals

An Afterword to "Bandits"

John Van Rys

As part of her work on the relationship between poetry and practical ecology, Deborah C. Bowen developed, in 2018, a series of leaflets that combine poems and images, as well as listing resources readers might use to put ecological principles into action. One leaflet focuses on human care for wild creatures and includes on its cover a photograph of raccoons, taken by her husband John of a family perched in a tree in their backyard. These are undoubtedly urban cousins to the raccoons in my story. Taken together, this leaflet and "Bandits" direct me towards my true subject: the complicated relationship between us and the other creatures, including our responsibilities to them. What is Shakespeare getting at when he has Hamlet describe humans as the "paragon of animals" (*Hamlet*, in *Works*, 1951, Act 2, Scene 2, l. 305)? Might this concept offer a clue to understanding the ecological territory of creature care?

Hamlet's phrase comes at the climactic moment in his reflection on the question "What a piece of work is man?" in Act 2, scene 2. After describing humans as the "paragon of animals," the model and perfect example of creaturely excellence, Hamlet turns away from his litany of praise, likening humans instead to a "quintessence of dust," devoid of meaning and purpose. Unless we can make some progress on this existential question of human identity, we have little hope of entering into a right relationship with the other animals.

The omnipresence of animals in human life and human cultural expression from earliest prehistory and across all cultures—in myths, in legends, in daily life, in literature, in art—itself shows the centrality of this relationship, and our need to address it. The extent of this task is suggested by a piece of writing advice offered by Joy Williams ("How to Write a Short Story," 2016). In her list of eight essential attributes of the short story, number four is "An animal within to give its blessing." Simply the presence of an animal in a story is a form of blessing, a term that suggests the possible religious significance of nonhuman creatures, in both life and literature.

My area of study is Canadian literature, and given the size and diverse geography of Canada, this canon of written work is rich territory for exploring the animal-human connection. One of Canada's best-known authors, Margaret Atwood, is also a champion of ecological issues, particularly in her recent dystopian trilogy *Oryx and Crake*, *The Year of the Flood*, and *MaddAddam*. Two of

DOI: 10.4324/9781003366744-9

Atwood's early works affected me profoundly as an undergraduate student: the poem "The animals in that country," first published in 1968, and her 1972 novel *Surfacing*. The poem sets up a sharp contrast between two worldviews regarding the human-animal relationship. In the first two-thirds of the poem, the speaker elaborates these opening lines: "In that country the animals/have the faces of people" (Atwood, *Selected Poems*, 1990, 40). Referencing cats, foxes, bulls, and wolves, the poem describes a cultured treatment of animals, something clearly Western or European, in which animals play roles that are "ceremonial," where they are "run/politely to earth" or "embroidered with blood." Then in the following lines, the speaker turns to an opposite perception of animals in a distinctly different environment: "In this country the animals/have the faces of/animals." The human presence here is reduced to "car headlights" in which "their eyes/flash once," then disappear. Animal deaths "are not elegant" in this wilderness space. To drive home the difference, Atwood closes the poem with "They have the faces of/no-one." Atwood's distinction here suggests competing understandings of, attitudes towards, and experiences of animals, ones that transcend particular geographies or cultures. On the one hand is domestication, control, and an interweaving of creatures into the fabric of human culture; on the other is an almost unknowable otherness, a wildness that escapes human view.

Surfacing explores this nature-culture relationship in narrative form. One image in particular offers a haunting symbol of human interaction with animals: during their trek, the narrator, her boyfriend Joe, and their friends, David and Anna, come across a Great Blue Heron—killed, decaying, hung upside down from a tree (1972, 115–117). With its wings spread, this bird clearly suggests an upside-down crucifixion in the manner of St. Peter, with disturbing latter-day overtones of lynching. David and Joe, who are making a postmodern pastiche of a movie, "shoot" the bird in film, symbolically killing it a second time, making "art" of it that echoes the cultured treatment of animals in the opening of "The animals in that country." Through this image, along with other narrative threads, Atwood starkly confronts the reader with questions of what might be called human inhumanity, of the possible meaning of the deaths of animals in relation to a kind of soul sickness.

Long haunted by Atwood's images, I have also experienced the human-animal dynamic up close. Although I had an urban and suburban childhood and youth, one formative experience for me was working on a dairy farm for a summer when I was 17; later on in adult life, my family and I took up hobby farming. We've been at it for more than 20 years, first in Iowa and then in Ontario, near Hamilton. This hobby-farm life has positioned us at the intersection of cultivated and wild nature, domesticated and untamed animals. We have raised and cared for horses and chickens and ducks, dogs and cats, even a ball python; we have also cohabited with wild creatures, from raccoons to skunks, possums to rabbits, deer to coyotes and wild turkeys. I've come to see this hobby-farm space as located between the two extremes Atwood articulates in her poem and novel, a space where tameness and wildness meet and need to be negotiated, and if possible reconciled, where my own and possibly all people's relationships to our fellow creatures need to be plumbed and lived out ethically. In this, I'm both inspired and chastened by a remark made by Wendell Berry in "Poetry and Place":

> The dichotomy of wildness and domesticity is a matter more complex and difficult for, unlike good and evil, wildness and domesticity, from the human point of view, are not entirely separable except as concepts and moreover are dependent upon one another. They are not interchangeable; they cannot *be* each other, but they cannot get along without each other. They are different, that is, not like good and evil, but like male and female; they are different, but not divided.
>
> (Berry, *Standing*, 1983, 192)

Berry's reframing of wildness and domesticity outside a good-evil binary hints at the complexity of a religious interpretation of the relationship among creatures. Put simply, what does the Bible claim about the relationship? Immediately, we're in the rough terrain of the cultural mandate and how "dominion" has been interpreted and practised for centuries. We're enmeshed in passages about animal husbandry and animal sacrifice that raise contemporary questions around food justice, sustainable agriculture, the ethical care of livestock, vegetarianism, and veganism; questions also involved in debates around climate change, speciesism, and more. Animals are everywhere in the Bible: the creation story, with days five and six dedicated to the making of all creatures, humans included; Satan's temptation in the likeness of a serpent; Cain's killing of Abel, a keeper of flocks whose sacrifice of some of his firstborn animals finds favour with God; Noah and the flood, and the covenant made with all creatures in Genesis 9; Daniel in the lion's den; Jonah preserved from drowning in the belly of the great fish. Consider the prevalence throughout of shepherds and sheep—literal, metaphoric, and symbolic. Or consider in the Sermon on the Mount Jesus's lesson about the birds of the air who are fed by God, though they neither sow nor harvest nor store away in barns.

Others have written on the theology of animals (for example, Andrew Linzey, *Creatures of the Same God: Explorations in Animal Theology*); my own humble point of departure is the first verse of Psalm 24: "The earth *is* the LORD'S, and the fulness thereof; the world, and they that dwell therein" (KJV). In my mind, this proclamation is not about God's "ownership" of this world but the world's createdness, its oneness with and value to Him. Moreover, this oneness and inherent value apply to everything within the earth, captured by that wonderful word "fulness" and the phrase "they that dwell therein"—who are "they," but all the creatures referenced in Genesis? The natural response to this beautiful proclamation should be for us to feel the utmost love, respect, and awe for our fellow creatures, for they are of God; they should be included in an ethic that calls for love of God with all one's being and for love of neighbour as oneself, all creatures being our neighbours on this planet. As Wendell Berry comments in "Economy and Pleasure,"

> these are the pleasures that we take in our own lives, our own wakefulness in this world, and in the company of other creatures—pleasures innate in the

> Creation and in our own good work. It is in these pleasures that we possess the likeness to God that is spoken of in Genesis.
>
> (Berry, *What*, 1990, 138)

From the first book of the Bible, he turns to the last, referencing Revelation 4:11, "Thou art worthy, O Lord, to receive glory and honour and power: for thou hast created all things, and for thy pleasure they are and were created" (KJV). Meditating on this verse, Berry suggests that

> our truest and profoundest religious experience may be the simple, unasking pleasure in the existence of other creatures that *is* possible to humans. It suggests that God's pleasure in all things must be respected by us in our use of things, and even in our displeasure in some things.
>
> (Berry, *What*, 1990, 139)

Such respect is essential to being the paragon of animals, whether one takes pleasure in chickens or is displeased by raccoons.

Religiously oriented writers know how far short of this ideal we fall, how much damage we have already done to the Earth and all who dwell in it. In her essay "Wilderness," Marilynne Robinson emphasizes the large scale of this damage. "Environmentalism," she writes, "poses stark issues of survival, for humankind and for all those other tribes of creatures over which we have exercised our onerous dominion. Even undiscovered species feel the effects of our stewardship." She then declares, "What a thing is man" (Robinson, *Death of Adam*, 1998, 245), echoing not only Hamlet's question but the opening line of Edward Taylor's "Meditation 38" (Taylor, "Meditation," 2012, 295). Referencing the creation story, she adds, "The oldest anecdotes from which we know ourselves as human, the stories of Genesis, make it clear that our defects are sufficient to bring the whole world down. An astonishing intuition, an astonishing fact" (Robinson, *Death of Adam*, 1998, 245). Our tragedy is in how far short we fall of being truly the paragon of animals—the paragon that should be the exemplar of God and God's joy in creation has also the power to devastate, even destroy that creation.

Wendell Berry and Mary Oliver echo Robinson's conclusions about the global scale of the problem and its source in humans, but they also effectively elaborate it at local and individual levels. In "Damage," the opening essay of *What Are People For?*, Berry tells the story of his failed attempt to create a pond for his cattle. Even with expert input and care, his effort harms not only the land but also his own spirit. "Until that wound in the hillside, my place, is healed," he confesses, "there will be something impaired in my mind. My peace is damaged. I will not be able to forget it" (Berry, *What*, 1990, 6). In his essay "People, Land, and Community," he articulates this problem as one of human arrogance, of a kind of folly rooted in an ignorance that is often revealed to us paradoxically in our acquisition of knowledge (Berry, *Standing*, 1983, 65). Here Berry argues that our machines and our tools are not solutions to problems but problems in and of themselves: "we are humans—which means that we not only *have* problems but *are* problems" (66–67).

In the opening to her essay "Sister Turtle," poet Mary Oliver shares her particular struggle over eating meat. She describes this issue as "a continually interesting subject of deep ambiguity," given her own devotion to Nature, as clearly evidenced in her poetry. However, she says, "to consider Nature without this appetite—this other-creature-consuming appetite—is to look with shut eyes upon the miraculous interchange that makes things work, that causes one thing to nurture another, that creates the future out of the past" (Oliver, *Upstream*, 2016, 49). She wishes to acknowledge this complexity, to recognize our participation in creaturely life-and-death dynamics, yet in the revealing passage that follows she expresses the difficult, troubling burden that goes with it:

> Still, in my personal life, I am often stricken with a wish to be *beyond all that.* I am burdened with anxiety. Anxiety for the lamb with his bitter future, anxiety for my own body, and, not least, anxiety for my own soul. You can fool a lot of yourself but you can't fool the soul. That worrier.
>
> (49–50)

What Oliver articulates here, and what she shares with Berry and Robinson, is a notion that to be the true paragon of animals, humans must make the human-animal encounter a soulful experience in all the senses of that word—one characterized by profundity, by depth of feeling, by joy mixed with sorrow. Such soulful being is filled with particular challenges and obligations.

As the paragon of animals, our task is to bring about healing, to establish harmony, to seek the flourishing of all creatures, not just our own peace. Indeed, human flourishing is impossible without the flourishing of all creatures. In seeking such flourishing, we would restore a right relationship to the highest degree possible. We would honour the value and significance of each creature in relation to ourselves and our Creator. In the opening sentence of "Healing," his aphoristic essay that functions as a companion to "Damage," Berry proclaims that "the grace that is the health of creatures can only be held in common" (Berry, *What*, 1990, 9). This is soulful living, living in which human culture functions harmoniously within Nature. "The made order," he argues, "must seek the given order, and find its place in it" (12). Specifically, "the field must remember the forest, the town must remember the field, so that the wheel of life will turn, and the dying be met with the newborn" (12).

Berry's vision is echoed in Robinson's conclusion to "Wilderness," where she argues that our ecological problems are essentially failures of civilization, not simply an inability to preserve wilderness. The solution is to be found in transforming civilization, in bringing it into a right relationship with Nature:

> I think we are desperately in need of a new, chastened, self-distrusting vision of the world, an austere vision. . . . Every environmental problem is a human problem. Civilization is the ecology being lost. We can do nothing that matters if we cannot encourage its rehabilitation. Wilderness has for a long time figured as an escape from civilization, and a judgment upon it. I think we

> must surrender the idea of wilderness, accept the fact that the consequences of human presence in the world are universal and ineluctable, and invest our care and hope in civilization, since to do otherwise risks repeating the terrible pattern of enmity against ourselves, which is truly the epitome and paradigm of all the living world's most grievous sorrows.
>
> (Robinson, *Death of Adam*, 1998, 253–254)

Paradoxically, to truly become the paragon of animals, to be rightly civilized, requires the defeat of a misplaced pride by a proper humility. Pride here takes on more than its metaphysical, spiritual, and ethical meanings: it becomes a matter of creation care, of the survival of creatures and creation. "Humility is the prize of the leaf-world," says Mary Oliver, "Vain-glory is the bane of us, the humans" (Oliver, *Upstream*, 2016, 7). Defeating this pride is a prerequisite for becoming a true creature. Robinson has eloquently argued in her essays for the exceptional nature of humans, in their ability to feel wonder at the creation. This wonder directs us into a right relationship with God and all his creatures, one in which humans can function as true paragons—as worthy examples of God's love for the creation. Berry locates us within time, within concentric circles, including "the larger circle of all creatures, passing in and out of life, who move also in a dance, to a music so subtle and vast that no ear hears it except in fragments" (Berry, *What*, 1990, 10). He concludes that "the more coherent one becomes within oneself as a creature, the more fully one enters into the communion of all creatures" (11).

Berry also recognizes that achieving this creaturely communion, what he calls elsewhere "an understanding and acceptance of the human place in the order of Creation—a proper humility" (Berry, *Standing*, 1983, 71), is no simple task. In "People, Land, and Community," he argues that we must realize we act daily in the face of insufficient information, being guided by love, good character, and proper culture (Berry, *Standing*, 1983, 67). In "Poetry and Place," he proposes that we must find ourselves in the coherence of nature rather than lose ourselves seeking to destroy it; that coherence we must accept "humbly as a gift at once natural and divine, and [we] must be materially and practically responsible for it" (Berry, *Standing*, 1983, 174). He couches this thinking within the tradition of the Great Chain of Being, positing that a recognition of creaturely difference, a respect for all life, and a proper human humility are the key to flourishing:

> . . . if one loves the Creator, one must love His creatures which, after all, live by His will and by sharing His life. When people keep their proper place in the Chain, the connections are unbroken from the microorganisms to God. When humans abdicate their proper place, either by pride or self-debasement, they blaspheme God and brutalize nature.
>
> (Berry, *Standing*, 1983, 183)

This vision brings us back to tragic plays such as *Hamlet* and humble comic stories such as "Bandits." What does literature have to offer here? Beyond blessing our stories with the presence of nonhuman animals, what can imaginative writers bring to this challenge? Put simply, their art: an art grounded in an understanding of our true creatureliness, an exploration of the challenges that condition brings, and a vision for what it might be. Such an art would be neither naively utopian nor loud, angry, and didactic. In "The Pleasures of Eating," Berry suggests that "farming, animal husbandry, horticulture, and gardening, at their best, are complex and comely arts" (Berry, *What*, 1990, 151). Literary arts sensitive to this vision of creatures can be a form of such husbandry, complex and comely in parallel ways. To borrow another thought from Berry, offered in "Poetry and Marriage," writers bring their baffled minds to this work, as "the impeded stream is the one that sings" (Berry, *Standing*, 1983, 97).

My own baffled mind has thus far for the most part found its stream to be the waters of comedy, the rivulets of humour. Exactly why, I'm not sure. However, I do know that what comedy does offer through its jokes and incongruities, its laughter and its transgressions, is a proper and productive form of humiliation, of bringing people down to earth into a more fitting understanding of what it means to be the paragon of animals. I believe that such laughter is "care of the soul" in the sense that Thomas Moore describes it. Towards the end of his book of that title, he reflects on our connection to "the soul of the world, *anima mundi*" (Moore, *Care*, 1992, 268). As part of that vision of how human souls are part of an ensouled universe, he includes our relatedness to the animal world, each animal expressing itself in its "remarkable individuality." "An animal," Moore states in language evocative of Gerard Manley Hopkins's instress and inscape, "reveals its soul in its striking appearance, in its life habits, and its style" (268). So true, when I think of chickens and raccoons—and of the creature that I am.

References

Atwood, Margaret. *Selected Poems: 1966–1984*. Don Mills, Ontario: Oxford University Press, 1990.

———. *Surfacing*. New Canadian Library. Toronto: McClelland and Stewart, 1972, 1990.

Berry, Wendell. *Standing by Words: Essays*. Berkeley, CA: Counterpoint, 1983.

———. *What Are People For?* New York: Farrar, Straus and Giroux, 1990.

Bowen, Deborah. 2018. *The Poetry and Ecology Project*. www.redeemer.ca/wp-content/uploads/Poetry-and-Ecology-Project.pdf.

Linzey, Andrew. *Creatures of the Same God: Explorations in Animal Theology*. New York: Lantern Publishing, 2009.

Moore, Thomas. *Care of the Soul*. New York: HarperCollins, 1992.

Oliver, Mary. *Upstream: Selected Essays*. New York: Penguin, 2016.

Robinson, Marilynne. *The Death of Adam: Essays on Modern Thought*. New York: Picador, 1998.

Shakespeare, William. "Hamlet." In *William Shakespeare: The Complete Works*, edited by Peter Alexander, 1028–1072. London and Glasgow: Collins, 1951, 1978.

Taylor, Edward. "Meditation 38 (First Series)." In *The Norton Anthology of American Literature: Volume A Beginnings to 1820*, edited by Nina Baym, 295–296. New York and London: W.W. Norton and Company, 2012.

Williams, Joy. "Joy Williams Explains How to Write a Short Story." Interview by Lincoln Mitchell. *Vice*, 18 July 2016. www.vice.com/en/article/avav9b/joy-williams-ninety-nine-stories-of-god-how-to-write-a-short-story.

10 From Grass to Galaxy

Alice Meynell's Poetic Wayfaring in the Meshwork of the World

Karen Dieleman

Alice Meynell (1847–1922) has been recognized in twenty-first-century criticism both as a poet committed to the Roman Catholic faith and as a poet and essayist invested in nature writing, but not yet substantially as a poet for whom theology and ecology were intricately intertwined—a connection this essay aims to establish. Although most scholars name Meynell's Roman Catholicism as a point of biography, only a few have given it literary-critical attention. In 2003, Kathleen Anderson briefly assessed how for Meynell, "The poet's ultimate task is to find God in her Son/poetry, and thereby, in herself" (Anderson, "Poetic Identity," 2003, 271) and Elizabeth F. Gray ably demonstrated the "fundamental coherence of Meynell's style, faith, and feminism" ("Alice Meynell," 2003, 160) in Meynell's Catholic-inflected poetry. Few analyses followed from these, though Isobel Armstrong pointed out in 2019 that Meynell's "new grammar of the sensoria [is] indivisibly bound up in her understanding of the core principles of Catholic belief—the importance of the incarnation, the living presence of the godhead in transubstantiation, the literal body and blood of Christ in the sacrament of the Eucharist" (Armstrong, *Victorian Poetry*, 2019, 463); and Lesa Scholl, in her 2020 monograph, *Hunger, Poetry and the Oxford Movement*, frequently associates Meynell's work with these Catholic beliefs. However, while Anderson, Gray, Armstrong, and Scholl all point to the centrality of Meynell's Catholic beliefs, sometimes also for her nature poems, their chief interests are in connecting Meynell's faith to her poetics, feminism, polity of the human, and/or social vision, not in establishing the fundamental coherence of Meynell's theology and ecology.

On Meynell as a nature writer, several scholars have written persuasively and insightfully, though not about the role of Meynell's theology in her view of the natural world. Linda Peterson argued in 2006 that in her earliest published collection, *Preludes* (1875), Meynell shows anxiety over possible influence by male Romantic poets who gendered Nature as Mother and that she later rejected these early poems for not offering a strong alternative to Romantic modes of nature poetry (Peterson, "Alice Meynell's 'Preludes,'" 2006). In a 2010 essay, Peterson turned to Meynell's nature essays, which she considers "more literary and ecological" than Grant Allen's (Peterson, "Writing Nature," 2010, 81), but she references Meynell's Catholic faith in a single sentence. Two subsequent and rich essays on Meynell's nature writing followed. In 2011, Rachel O'Connell wrote perceptively that "the

DOI: 10.4324/9781003366744-10

ever-present danger in the world of Meynell's essays is the arrogance and acquisitiveness of impulses towards mastery, [impulses that desecrate] the right order of a world that provides certain natural freedoms" (O'Connell, "Significant Negatives," 2011, 66), but despite an interest in Meynell's ethic of love for the other, O'Connell does not mention Meynell's faith. Similarly, Patricia Murphy, who devotes a chapter to Meynell in her *Reconceiving Nature: Ecofeminism in Late Victorian Women's Poetry* (2019), gives as the "most important concept in Meynell's ecopoetics" "the conviction that nature is a vibrant entity on its own terms. . . . The poet effectually becomes an instrument of nature, which always remains a robust presence as an inspirational creative body" (129), but Murphy also does not mention Meynell's faith or its import for her thinking about nature.

In contrast to the general separation of these two lines of inquiry into Meynell's writing, a 2013 essay by Hilary Fraser centralizes the relationship between Meynell's Catholic Christian faith and her understanding of the natural world. Comparing Meynell with Gerard Manley Hopkins in her study of visuality in these poets' natural theology, Fraser writes:

> The Incarnation was similarly central to Meynell's faith, and defined her own dynamic sensory and emotional engagement with the phenomenal world. . . . For Meynell, as for Hopkins, the continuing 'real' presence of Christ in the world authorized a newly revitalized Christian hermeneutic by which the aesthetics of landscape could be read in an interactive way. . . . Hers is, like Hopkins's, a sacramental universe that is revealed to be constituted by its fundamental incarnate relatedness.
>
> (Fraser, "Aesthetics," 2013, 95)

Fraser briefly illustrates her thesis with reference to two Meynell essays and three poems. Citing an earlier work by Fraser that makes the same point, Mark Knight and Emma Mason, in their overview of nineteenth-century religion and literature (2006), also focus on Meynell's essays (and, briefly, one poem) to trace the centrality of Meynell's Catholicism to her thinking about the dynamic relation between art and nature (Knight and Mason, *Nineteenth-Century Religion and Literature*, 2006, 200–202).

I offer the present article as a development of Fraser's and Knight and Mason's recognitions and as the first full-length chapter or article reading Meynell's poetry as integrated theology-ecology. I do so by applying to Meynell's poetry ideas from sociologist Tim Ingold, as refracted through theologian Norman Wirzba. Ingold, in *The Life of Lines* (2015), considers both sociality and ecology as a "holding on to one another" made possible by the putting out of lines that then entangle to form a meshwork (5) that is "boundless and ever-extending" (11)—in ecology, sometimes visible in animal or plant trails that "follow no consistent direction [but] are continually responsive to environmental variations" (82), and, in sociality, following "lines of speech, song, story and handwriting" (87). Differing from a network, which has planned nodes of connection, a meshwork made of such lines is a tangle, an entanglement that emerges rather than is made. Further, in such a meshwork, things are not "objects" that "exist" but "things" that "occur"—"that carry on along their lines," existing in the world "not as nouns, but as verbs—as goings-on"—and including "such meteorological phenomena as sunshine, rain and wind" (16). Ingold details two ways of human movement through this meshwork: navigating and wayfaring. A navigator seeks to move "point-to-point," "by calculation," with a "determination to get from A to B within the space of possibilities offered." Objects encountered in the way are therefore barriers to the destination. But for the wayfarer, "every destination is by the way; his path runs always in between." His only objective is to keep going, but as he goes, he must "pay attention to things, and adjust his gait accordingly." Such movement is "not so much intentional as *attentional*. It thrusts the follower into the presence of the real" (133).

Ingold offers no theological framework for his ideas, believing as he does that "We creatures are adrift. Launched upon the tides of history, we have to cling to things, hoping that the friction of our contact will somehow suffice to countervail the currents that would otherwise sweep us to oblivion" (3). But theologian Norman Wirzba draws Ingold's ideas into a Christian theological framework helpful for approaching Meynell's thought. In "Creation *through* Christ" (2018, italics original), Wirzba proposes Ingold's ideas of meshwork and wayfarer as useful for understanding the sort of place that creation is. Since God's creating is a "sign of a divine, hospitable intention for others to be"—an act of love which "desires above all else . . . for creatures to thrive and flourish" (38)—creation, as a meshwork, is nothing less than "the establishment of a spiritual and moral *topography* that alerts us to what the right ordering and empowerment of creatures looks like in the flesh. . . . [Creation] is about the *character* of the world and its proper *orientation*" (39–40, italics original). However, given human failing, "we do not know what things are or how we should engage them apart from Christ's work to love and heal everything" (40). Therefore, the "decisive key" to relating to created things so that they can flourish as intended in the creational meshwork is "to find Jesus and enter his *tropos* because creatures actually (and paradoxically) more fully realize themselves insofar as they participate in Christ's way of being. Christ's way of love, his being for others, is 'the way'" (42). Citing Colossians 1:16–17, that in Christ "all things in heaven and on earth were created" and that "all things have been created through him and for him," Wirzba roots the significance of the material world in the presence and activity of Christ at creation as "the eternal *Logos*" (36).

Though Wirzba does not say so, in its plenitude of prepositions relating Christ to the noumenal and phenomenal world, the Colossians text contributes well to the idea of creation as a spiritual and moral topography. This plenitude is augmented when different English translations of the Greek are juxtaposed: the three clauses of the Colossians text then state all things were created *in/by* Christ, that all things have been created *through/by/for/in* him, and that all things hold together *in/by* him. This prepositional matrix points to the comprehensiveness of Christ's relationship to all things. Christ has an originating, continuous, purposeful, sustaining, and dynamic relationship with everything that exists. The result is that all things function in relationship to each other as well as to their shared creator. This is what makes the created world a meshwork rather than a system; by virtue of the Christ *in-by-through-for* whom they were made, all created things are also *in-by-through-for*; they interact; they put out their lines; they happen. Moreover—and as we will see, this next recognition is particularly important to Meynell—this created meshwork receives further material and spiritual significance from the incarnation, whereby Christ, the Son of God, entered his own creation as a real human being named Jesus who lived and died on earth. Though—as both Protestant and Catholic Christians confess—he also rose from the dead and returned to heaven as God-Man, Catholics emphasize his continuing real presence in the world, most centrally in the mass (wherein the bread and wine become, in Catholic theology, Christ's body and blood anew—his Real presence), but also in the liturgy of the church, the natural world, human activity, and so on. Thus, for the Catholic Meynell, Christ's incarnated entry into the meshwork of creation brings all *in-by-through-for* things not merely into the presence of the real, as for Ingold and Wirzba, but into the presence of the Real.

Further, as the poetry reveals in different ways, Meynell imagines the incarnated Christ as being and moving within this meshwork as a wayfarer—indeed, as the ultimate Wayfarer. He is not passing through, point to point as a navigator moving past objects that merely exist, but he attends fully to all goings-on and extends his lines into the lives and actions of all by the way—which is to say, by the Way. For, on the basis of his eternal, creating, loving relationship to all things, Christ makes exactly that comprehensive claim, namely, that though he inhabited his creation, he is also the Way, the Truth, and the Life for all things in that creation (John 14:6). It is a claim Meynell acknowledges in a key poem that opens my discussion of her poetry in the following section, and is perhaps part of the reason she finds her idea of Christ as wayfarer so appealing. Set next to each other, the two concepts of Way and Wayfarer express the paradoxical power and humility of Christ in relation to all things; in their paradoxical richness, the two ideas convey her complex yet subtle ecotheology.

This ecotheology includes the conviction that Christ's way of being in creation also guides the poet's wayfaring through nature and language. In this thought, Meynell follows another Pauline concept, expressed in Romans 8, one that both aligns and distinguishes between humans and the rest of creation. Paul situates humans as part of the creation that is yearning for Christ's return in order to be freed from "bondage to decay and brought into the freedom and glory of the children of God" (8:21); he uses the same verb, "groan," to describe their response to the burden of brokenness. But by virtue of his writing, Paul also implies a distinction between human and nonhuman, namely, that the human alone can use language to "find Jesus and enter his *tropos*" (Wirzba, "Creation *Through* Christ," 2018, 42). This distinction allows the poet to wayfare like Christ in the meshwork of the world, bearing witness to its wonder and pain, imitating Christ's way of being in order to help things flourish as intended in God's creating act of love. The poet shares in but also draws attention to the spiritual topography of the world, inviting others to become attentive and responsive wayfarers, too.

In what follows, I aim to show that Meynell's theological understanding, particularly of Christ as wayfaring within the world, bears fundamentally—even comprehensively: from grass to galaxy—on her poetry about nature. Meynell considers the incarnated, crucified, and yet living Christ as continuing to inhabit or wait within or walk among the grain, grass, grapes, and galaxies of the world, devoted to their particularity. Christ's coming and going—his wayfaring among the elements of the world and his way-of-being for them—marks the world, such that each element of the world also has a particular way of being and becoming: to use Ingold's language, that of being a thing that occurs and not an object that exists. Included in this *tropos* is any person but especially the poet who goes as a wayfarer among things and not as a traveller past objects. Such a poet sees the true character of creation as a meshwork and, attentive and attuned to this spiritual topography, also wayfares through language towards the surprise often lurking in the ordinary.

Christ as Way and Wayfarer

Meynell opened her privately printed *Other Poems* (1896)—as well as her next collection, *Later Poems* (1902)—with "The Lady of the Lambs" (discussed in the next section) and two other lyrics that lead the way in this study because of their significance for grasping her motif of Christ as Way and Wayfarer: "I Am the Way," and "Via, et Veritas, et Vita." With these two titles, Meynell foregrounds Jesus's self-identification as the Way, the Truth, and the Life (John 14:6). The speaker of "I Am the Way" begins with what appears to be a simple confession that acknowledges this claim: "Thou art the Way." Readers might have expected this opening line to continue by following Jesus's own explication of this expression, as meaning "No one comes to the Father except through me" (John 14:6)—as Bishop Doane did in his well-known nineteenth-century hymn whose first stanza runs, "Thou art the Way; by Thee alone/From sin and death we flee;/And he who would the FATHER seek/must seek Him, LORD, by Thee" (Doane, "Thou Art the Way," 1861, ll.1–4). But Meynell's speaker does not follow this route of hymnic paraphrasing of Scripture. She is in no hurry to get to the Father in the poem itself. Rather, in three condensed quatrains of alternating bimetre and tetrametre lines, she considers what it means for herself, for Christ, and for poetry that Christ is Way and Wayfarer.

Thou art the Way.
Hadst Thou been nothing but the goal,
I cannot say
If Thou hadst ever met my soul.

I cannot see—
I, child of process—if there lies
An end for me,
Full of repose, full of replies.

I'll not reproach
The road that winds, my feet that err.
Access, approach
Art Thou, Time, Way, and Wayfarer.
(*Collected Poems*, 28)

In metre and meaning, these lines exhibit both progress and halt. In the first stanza, the speaker implies relief that Christ is the Way, for if he were only the end point, she is not sure she would have or could have stayed on the path long enough to reach him. Why not? Because, as the second stanza indicates, she is a "child of process," not a child of destinations: or, in Ingold's terms, she is a wayfarer, not a navigator. She "cannot see" whether there is an end, or what kind of end it might be. But the third stanza reveals that such uncertainty about ends does not alarm her, because her winding roads and erring feet do not nullify Christ as Way and Wayfarer. Her meandering—even when in error—cannot close off his Access. Indeed, this conviction that irregularity does not threaten the wayfarer who trusts the larger topography marks the poem's form as well: the poetic feet in each line are syllabically exact (unerringly alternating four and eight syllables), but the expected iambs are often on the move, shifting into trochees (for example, "Full of"; "Access") and spondees (for example, "I, child"; "Art Thou"; "Time, Way") or offering metrical options (for example, "Thou *art*" or "*Thou* art"). Expected grammatical functions of individual words also shift as those words entangle with their surroundings: the penultimate line initially presents as a pair of verbs whereby the speaker encourages her erring feet to access or approach; but "Art" as verb in the final line turns "Access" and "approach" into noun equivalents for Christ, akin to Time, Way, and Wayfarer. The poem's lines, that is, respond to their own environment. This wayfaring characteristic of the poem occurs between the stability of the "Way" in the opening and closing lines, a concept that Meynell explores further in the next poem (partnered in all her collections with "I Am the Way").

"Via, et Veritas, et Vita" is a single quatrain consisting of an exchange between two voices. An initial speaker accuses another, "You never attained to Him" (1), and the accused admits such might be the case, "If to attain/Be to abide" (1–2)—an equation that she implicitly and tonally does not grant. The first speaker then responds, apparently from the place of attainment (problematic in its lack of humility, its failure to be "child of process"), with a complaint about the length and difficulty of the journey, a complaint that doubles as self-praise for perseverance through suffering: "Endless the way, followed with how much pain!" (3). The quiet reply of the accused, however, is a corrective to this complaint about the hard way: "The way was He" (4). The first speaker, that is, thinks of Christ only as destination and, consequently, wishes the way to him had not been so difficult. The accused, however, makes clear that a critique of the journey is a critique of Christ himself, as well as a failure to perceive him according to his self-revelation as the Way, the Truth, and the Life. Instead, to attain to Christ is to be ever-walking with him along his own way. With its precise and powerful brevity about Christ as Way, this poem functions as a short excursus from "I Am the Way."

But to return to the third stanza of that preceding poem, the "child of process" concludes by identifying the Way also as the Wayfarer, and this identity clears her from possible guilt over her own non-linear journey. If, as Wirzba states, the "decisive key" to relating to created things is "to find Jesus and enter his tropos," then the poet's wayfaring—her being a child of process more than of goals—needs no reproach, for it patterns Christ's own wayfaring.

What Meynell means by Christ's being Wayfarer as well as Way emerges more fully across her poetry of the next decade, especially in "Unto Us a Son Is Given" (1902),[1] "The Fugitive" (1907),[2] "Christ in Portugal" (1911; later retitled "In Portugal, 1912"),[3] and "Christ and the Universe" (1911; later retitled "Christ in the Universe"). In these poems, the wayfaring Christ is more closely connected to the material as well as spiritual meshwork of the world. In the earliest of the four, Meynell opens and closes by emphasizing the perpetual coming of Christ as the "Infant of mankind" (3). In something more than a preached gospel of good tidings, Christ himself "comes with tidings and a song" (7) each year, because he is "Given, not lent,/And not withdrawn—once sent" (1–2).[4] "New every year,/New born and newly dear,/He comes" (5–7). Implied here is that the Christ who is not withdrawn but who has moments of recognized coming must be elsewhere in the world in the intervals between comings. Not that the incarnation repeats but, Meynell suggests, having come once, Christ somehow remains in the world. Though once a year, in the nativity season, he comes "with expected feet" (14) as "the little welcome Son" (4), he is actually always somewhere, wayfaring in the world in a more hidden fashion. Moreover, his perpetual coming as an infant can be understood by considering the pattern of the creation within which he walks, as the third and fourth stanzas of the poem indicate: "Even as the cold/Keen winter grows not old,/As childhood is so fresh, foreseen/And spring in the familiar green.//Sudden as sweet/Come the expected feet./All joy is young, and new all art,/And He, Too, Whom we have by heart" (9–16). That is, every year, seasons and childhoods have their birth, and so do joy and art: just so, he "Whom we have by heart" comes as the infant Son, new but not unexpected.

In other poems, Meynell considers how Christ's wayfaring in the world might take the hidden fashion of a fugitive or even a yet unknown form. The first of these—the fugitive—Meynell explores in "The Fugitive" and "In Portugal, 1912," two poems responding, also in their syntax and metre, to religious-political events on the European continent. "The Fugitive" represents the movement of the escaping fugitive in the sudden start of its first line (a prepositional phrase with multiple objects but no prefatory subject or verb) and in the varied movement of its feet through enjambed lines and trochaic starts to otherwise iambic lines. Its subject is Meynell's reflection on the words of a French publicist who declared, on the disestablishment of the Roman Catholic Church in France in 1905, "Nous avons chassé ce Jésus Christ" (We have chased out this Jesus Christ). Meynell responds to this triumphantly secular epigraph with a "Yes" that she immediately qualifies: yes, you have driven him out of your hearts, public conversations, and homes, but you have not, and never can, expel him from the grain and vineyard; this Jesus Christ, in his particularity, never will escape the meshwork of the field, for he is inextricably bound to sheaf and grape, the bread and wine that—in Meynell's Catholic theology—continually transform him into flesh and blood and make him ever present in the world, entangling his lines with those of crops, sun, soil.

Yes, from the ingrate heart, the street
Of garrulous tongue, the warm retreat

Within the village and the town;
Not from the lands where ripen brown
A thousand thousand hills of wheat;

Not from the long Burgundian line,
The Southward, sunward range of vine.
Hunted, He never will escape
The flesh, the blood, the sheaf, the grape,
That feed His man—the bread, the wine.
(*Collected Poems*, 46)

Elizabeth Gray rightly observes that these final lines indicate that for Meynell, "just as [Christ] has not avoided incarnation, neither will he avoid being killed, denatured, both tragically and triumphantly transformed into these Eucharistic emblems" (Gray, "Alice Meynell," 2003, 167), and Lesa Scholl observes "the anaphoric use of the definite article" in these same lines as "particulariz[ing] Christ's presence in the world" (Scholl, *Hunger*, 2020, 112). But intriguingly for my argument, Meynell does not centre her Eucharistic meaning around the verb "transform." She chooses "escape" (8) and thus harks back to her title, which identifies Christ as "the fugitive"—signifying, in the multiple connotations given by the *Oxford English Dictionary*, "one who flees or tries to escape from danger" (B.1.a), "one who quits or is banished from his country" (B.1.c), and "one who . . . moves from place to place; a vagabond, wanderer" (B.2)—or, to pull these together into the concept I am exploring: Christ, wayfarer by choice but also by necessity, driven out of public sight and hearing but not thereby out of the material world, where he lurks in the "thousand thousand hills of wheat" (5) and "Southward, sunward range of vine" (7), potentially everywhere in the meshwork of creation as the immanent Christ.

Christ's immanence in the meshwork of the world is also the subject of Meynell's three-quatrain poem in response to the revolution in Portugal that established the First Republic in 1910, with its severe oppression of the Roman Catholic Church. Though she does not use the word "fugitive" this time, the same idea is present as she explores the effort, but ultimately, the inability, of humans to chase Christ out of the material world: he hides, waits, is lonely, in ambush, lurks, is concealed in wheat and wine—but he is not absent. On the contrary, though unconsecrated, unsacrificed, and unbidden because mass is politically forbidden and ignored, he is yet abundantly present in field and town:

And will they cast the altars down,
Scatter the chalice, crush the bread?
In field, in village, and in town
He hides an unregarded head;

Waits in the corn-lands far and near,
Bright in His sun, dark in His frost,
Sweet in the vine, ripe in the ear—
Lonely unconsecrated Host.

In ambush at the merry board
The Victim lurks unsacrificed;
The mill conceals the harvest's Lord,
The wine-press holds the unbidden Christ.
(*Collected Poems*, 47)

Scholl notes that the greater "regularity of form" in this poem compared to "The Fugitive" suggests an "internal confidence" (Scholl, *Hunger*, 2020, 112) that she connects to an early reviewer's comment: "with beautiful poetic mastery, [the poem] reveal[s] to us through the light of revelation the wondrous integrity and harmony of God's entire universe" (112), which Scholl connects to Eucharist meaning (113). That is, even as a fugitive, the incarnated Christ holds all things together in *in-by-through-for* relationship with himself and the rest of the meshed world. For instance, in stanza two, not only do sun and frost belong to him ("His sun," "His frost"), but their characteristics become his—or perhaps they always were his—and Meynell suggests we misspeak when we say "bright sun" or "dark frost" or "sweet vine" or "ripe ear." In the poem, it is Christ, *in* the things that he has made, who is bright, dark, sweet, and ripe as he waits in the corn-lands. Thus the futility of casting down altars and scattering the Eucharistic chalice and bread: these are only a few of the things that signify Christ in the world. Indeed, he is present not only in the natural world but also, as stanza three indicates, in human activity: at the ordinary table, in the flour mill, and in the wine-press. In sum, Meynell takes contemporary events that were oppressive for the Roman Catholic Church in parts of Europe and underscores, in "The Fugitive" and "In Portugal, 1912," the impossibility of separating Christ from what he has made and entered.

Meynell in 1911 was also imagining Christ in his creation as more than a fugitive wayfarer within fields and vineyards. In "Christ and the Universe" (later "Christ in the Universe"—a swap of preposition for conjunction that aligns better with her concept of integrated relationship)[5]—she includes terrestrial and celestial spheres to consider Christ's relationship to the entire meshwork. The poem is perhaps most known for its startling closing stanzas, which call on the soul to be prepared to hear, at some point "in the eternities" (21) through a "million alien Gospels" (23), that Christ not only was incarnated as a man on earth but also appeared in myriad other forms to the stars of the universe: "O, be prepared, my soul!/To read the inconceivable, to scan/The million forms of God those stars unroll/When, in our turn, we show to them a Man" (25–28). In this theological speculation, Meynell again casts Christ as the one not bound to the premade path. Of his movements on earth, Meynell writes, in a stanza that combines the notions of Christ as Wayfarer and Way, "Of His earth-visiting feet/None [i.e. no star] knows the secret, cherished, perilous,/The terrible, shamefast, frightened, whispered, sweet,/Heart-shattering secret of His way with us" (9–12). Of his movements in the universe, Meynell writes, "Nor, in our little day,/May His devices with the heavens be guessed,/His pilgrimage to thread the Milky Way" (17–19) or how "He trod the Pleiades, the Lyre, the Bear" (24). In these depictions, Christ is not fugitive nor is he aimlessly wandering in or passing through his creation to some other destination. Rather, he engages with purpose all that he seeks or encounters. He follows and attends to the boundless and ever-extending meshwork because it is *his* meshwork, with which he has "His way" and "His devices" and to which he gives "bestowals."

Across these poems, Meynell demonstrates that, from grass to galaxy, Christ relates materially and spiritually to what he has made. I explore next how, for Meynell, Christ's incarnated relationality as Way and Wayfarer enables and inspires the poet towards a similar attentiveness to the material and spiritual significance of created things.

The Poet as Wayfarer in the Meshwork of the World

In her 1913 *Collected Poems*, Meynell moved "I am the Way" and "Via et Veritas et Vita" out of second and third place in her sequence but retained "The Shepherdess" (initially "Lady of the Lambs") as lead poem. A fair amount has been written about this poem, often critically, because of the moral constraints it places on the shepherdess of thought: "Her flocks are thoughts. She keeps them white" (3); "She is so circumspect and right;/She has her soul to keep" (15–16).[6] But if we recall that Meynell twice placed the poem next to "I Am the Way"—with its "I'll not reproach/The road that winds, my feet that err" (9–10)—it does not seem she was overly concerned about possible critiques of her roving mind. While I do not want to reduce its complex meanings, I therefore take the poem as at least a gesture towards the idea of the poet as wayfarer. In it, Meynell depicts the poet as one who "walks" (1, 11, 17) and "roams" (7) hills and valleys, and whose thoughts simultaneously "run and leap" (14). Though this shepherdess of thoughts does safeguard them ("She holds her little thoughts in sight" [13]), yet by virtue of being a shepherdess, she is a rover as well as keeper. Together with her sheep/thoughts, she is nomadic: walking, feeding (5), and sleeping (6) on "the fragrant height" (5), "maternal hills" (7), and "Dark valleys" (8) around her. There are faint associations here between Christ and the shepherdess in their way of being in the natural world (associations strengthened by the poem's echo of Psalm 23). Like Christ, the shepherdess-poet goes into her surrounding world, attentive to its moral topography, to the significance of its heights and valleys and stars.

But Meynell had more—and better—ways of imagining the poet as wayfarer in the world, seeing the *in-by-through-for* significance of all things. Murphy, in her study of mainly pre-1900 Meynell poems, helpfully distinguishes between those that critique the poet for whom nature is "merely a vehicle providing access to a desired objective" and those "in which closeness with the nonhuman world leads the admirable nature poet towards worthy work" (Murphy, *Reconceiving Nature*, 2019, 150)—a distinction, I note, akin to Ingold's between navigator and wayfarer. In Meynell's post-1900 work, poems of the latter type often depict the close encounter of the poet with the created world as an embrace that releases. In these poems, Meynell frequently uses the word "close" as a verb, following definition 11 of the *Oxford English Dictionary*: "To come close together in contact or union; to join, unite, combine, coalesce, meet in a common centre." For her, when two entities in the meshwork willingly close or embrace, neither one is confined or restricted; rather, creative speech or thought is activated. Closeness enables a powerful mutuality for those who are attuned to each other's significance in the meshwork. For instance, "West Wind in Winter," first published in 1902, analogizes wind and poet-inspiration as forces that rush upon the awareness of those who go out to meet them. The poet-speaker seeks out the wind she knows well ("I know his note, his lay,/His colour and his morning-rose" [10–11]) in order to "confess his day" (12); equally true, the wind seeks out the poet, eager to activate possibilities. They are both agents in the encounter and urgent (in "haste") for intimacy:

West Wind! I would not miss
His sudden tryst: the long, the new
Surprises of his kiss.

Vigilant, I make haste to close
With him who comes my way.
. . .

. . . at dawn I hark
His call; at morn I meet
His haste around the tossing park[.]
(*Collected Poems*, 23–24,
lines 4–8, 13–15)

Perhaps the wayfarer motif construed so far—the attentive wanderer moving by the way—is not readily apparent in a poem expressing so much haste and turbulence, but the poet who chooses to be buffeted by wind is choosing to be "responsive to environmental variations" (Ingold, *Life of Lines*, 2015, 82), as she puts out her lines. She does not close with the wind in battle (though "close" can have that meaning) but in order to "confess" it truly. The ethos of the poem is exhilaration at such intimate knowledge of another thing in the meshed world. In the poem's final stanza, the speaker associates her wind-tryst with the rush of poetic inspiration that she feels and knows intimately and also confesses whenever the close encounter happens. This close encounter she finds, not as with the wind, in street or park, but united in flight as with a rushing bird:

So too, do I confess
My poet when he sings.
He rushes on my mortal guess
With his immortal things.
I feel, I know him. On I press—
He finds me 'twixt his wings.
(19–24)

Wind, poet, bird: these three entities weave through many of Meynell's poems in ever-evolving relationship: "Her flocks are thoughts" ("The Shepherdess," 3), a "flock of winds came winging . . ./Strong birds" ("The Roaring Frost," 1, 2), "bring back/Thy flocks of fancies" ("The Fold," 2–3), a "flock of bells take flight,/. . . A verse of bells takes wing" ("Chimes," 3, 11),"Flocks of the memories of the day draw near" ("At

Night," 3). Always, each entity—whether wind, or poet, or bird . . . or water, or leaf, or grass—becomes more fully itself when it closes in embrace with another.

While in the poem just considered, Meynell's poet is *associated* with or attentive to elemental forces, elsewhere the Meynellean poet actually *is* an elemental force, entering creation de-carnated, so to speak—imitating Christ's intimacy with the created world but necessarily in the reverse manner with regard to form. That is, in these poems, the poet divests herself of human form, becoming wind or water in order to wayfare more intimately within the meshwork—or to close with it still more. A minor instance of this fusion occurs in "The Watershed" (first publication, 1902; *Collected Poems*, 18), where the poet describes her journey across the continental divide from Munich to Verona in terms of her relationship to the waters. She is separate from the rivers flowing north—"I met, opposed, withstood/The northward rivers on their way" (9–10)—but she becomes one with those rivers flowing south: "Oh to see/Of all the southward brooks the first!/. . ./down a thousand vales I dropped,/I flowed to Italy" (16–17, 20–21). The switch is from navigator to wayfarer: from feeling the force of nature as a barrier to resist, to experiencing it as a co-element to move with and in. A more extended example occurs in "The Two Poets," first included in the latter pages of 1902 *Later Poems* and then, significantly, moved to second place—right after "The Shepherdess"—in *Collected Poems*. The relocation signals the increased importance to Meynell of this reimagining of the poet as articulated nature. The poem represents the two poets of its title as a west wind and a beech tree, inarticulate until they close with each other, after which they each first claim, then yield primary agency to the other:

> Whose is the speech
> That moves the voices of this lonely beech?
> Out of the long west did this wild wind come—
> O strong and silent! And the tree was dumb,
> Ready and dumb, until
> The dumb gale struck it on the darkened hill.
>
> Two memories,
> Two powers, two promises, two silences
> Closed in this cry, closed in these thousand leaves
> Articulate. This sudden hour retrieves
> The purpose of the past,
> Separate, apart—embraced, embraced at last.
>
> "Whose is the word?
> Is it I that spake? Is it thou? Is it I that heard?"
> "Thine earth was solitary, yet I found thee!"
> "Thy sky was pathless, but I caught, I bound thee,
> Thou visitant divine."
> "O thou my Voice, the word was thine." "Was thine."
> (*Collected Poems*, 14)

In the act of closing or embracing in this poem, the tree-poet catches and binds the wind, not to constrain it but to activate its purpose, turning its memory, power, promise, and silence into voice. Likewise, the wind-poet closes with the single tree and activates its purpose, turning its memory, power, promise, and silence into voice. The two distinct and dumb things unite in a cry nuanced as a "thousand leaves/Articulate" (9–10). In Ingold's terms, the wind and beech each extend their lines until they mesh and give audible testimony to their being things that occur. In their embrace, they achieve a "holding on to one another" (Ingold, *Life of Lines*, 2015, 5) that liberates each to achieve its full self. The poem concludes with the two elements deferring to each other—or perhaps arguing with each other—as to who actually gave voice. But the question that opens the poem—"Whose is the speech?"—and that is reiterated in the final stanza—"Whose is the word?"—is not resolved in favour of beech or wind. Instead, a different resonance emerges across certain words (word, earth, sky, divine, Voice) in the final stanza: a faint suggestion that what enables any created thing to break out articulate is its embrace with the creative Word, the Logos from the beginning, the immanent Wayfarer.

Whereas in "West Wind in Winter" and "The Two Poets," Meynell excavates the positive valence of "close" by linking it with tryst, kiss, and embrace, elsewhere she explores how one entity can close with another to lock it rather than embrace it, thus stifling rather than releasing its voice—in Ingold's terms, forcing it to exist rather than occur, disallowing its lines to wayfare. That type of closing with another, and its effect on the poet, appears in "A Poet's Wife," also first published in *Later Poems* (1902). In this lyric, a poet's wife observes how the poet becomes like "a tract of ocean locked inland" (1–2) whenever he "closest with the kiss of her/Who locks thee to her breast" (11–12). Since the poet is rightly "a sea of thoughts" (9), he becomes restless when so locked; he begins to "stir" (9), desiring to be more like the "very sea" (3) that freely moves about. Interestingly, in *Collected Poems*, Meynell replaced "stir" with "err," flipping the gender dynamic of the poem from a wife's recognition of her poet-husband's desire to escape her lock to a wife's rebuke to her poet-husband who chooses the locked embrace rather than go out, like the sea, to meet the night. In this second version, the poet's wife urges the mistaken poet to follow the example of the ever-moving, ever-seeking ocean waters, even at the cost of loneliness and hardship:

I saw a tract of ocean locked inland,
        Within a field's embrace—
The very sea! afar it fled the strand,
        And gave the seasons chase,
And met the night alone, the tempest spanned,
        Saw sunrise face to face.

O Poet, more than ocean, lonelier!
        In inaccessible rest
And storm remote, thou, sea of thoughts, dost stir [err]
        Scattered through east to west—

Now, while thou closest with the kiss of her
Who locks thee to her breast.
(*Later Poems*, 35; *Collected Poems*, 41)

It is fascinating that Meynell replaced "stir" with "err" in 1913—*after* she had written many of the poems discussed earlier in the "Christ as Wayfarer" section. It is as if, having built a clearer vision for herself of Christ as wayfarer, she saw how a single word shift in an existing poem could more closely align the human poet with that Christ-motif. A poet locked to one location—whether literally or metaphorically—cannot become part of the goings-on of things and will ultimately fail to understand the meshwork, much less make new lines or meanings to occur within it. The metaphors of this poem are wilder, the tone fiercer, than in "The Shepherdess," yet the basic thought is the same: to find creative power, the poet must wayfare beyond the human home or lover into the created world, embracing each entity but locking (to) none.

Such a poet can even observe the spiritual and moral topography of creation in towns, where it may go unnoticed by those too absorbed in their own lives to grasp the intricate meshwork around them. For reasons of space, I am brief here, but in "A Dead Harvest" (*Collected Poems*, 17), Meynell laments that unlike "rows of hay" (3) raked into "a stack" (11) for life-giving purposes, the dead leaves of Kensington Gardens become a "futile crop" (11) raked into a "pyre" (12). Worse, "So go the town's lives on the breeze,/Even as the sheddings of the trees;/Bosom nor barn is filled with these" (13–15). Such a "futile crop" contrasts sharply with the Christ-inhabited sheaf and grape of "The Fugitive" and suggests that while some try actively to hunt out the spiritual from the meshwork of creation, others simply ignore its presence. But the poet notices, and so may anyone who lives in a town, even if the experience of the spiritual topography of the world is more muted there. Meynell conveys this sense in "November Blue" (*Collected Poems*, 16), where the city of London has blurred the blue from its skies, such that "hooded in an earthly brown,/Unheaven'd the city lies" (3–4); when the "gold and silver lamps" (9) shine through winter mists, however, a substitute sky can occur, as through the effects of light and water: "Blue comes to earth, it walks the street,/It dyes the wide air through;/A mimic sky about their feet,/The throng go crowned with blue" (13–16). As in "A Dead Harvest," the throng seems not to notice its own crowning this way, but the observant poet, wayfaring in towns as well as fields and forests, bears witness to the glories of gold and blue wherever they occur in the meshwork of the world.

The Poet as Wayfarer Through Language

Such a poet also wayfares through language towards the surprise often lurking in the ordinary. As noted earlier, in describing things as things and not objects, Ingold asserts that we then "admit them into the world not as nouns but as verbs" (Ingold, *Life of Lines*, 2015, 116). For Ingold, this reconceptualization excludes any recognition of levels of transcendence (152); everything necessarily becomes "an immanent life lived mid-stream, in the in-between, where there are no subjects, no objects, no subject-object hybrids; only verbs" (152). Meynell, obviously, would not follow Ingold in refusing to recognize transcendence, but her focus on Christ's immanence in creation perhaps fuelled a similar interest in verbs as indicators of abundant, relational activity. As Knight and Mason observe,

> she remained confident that language was capable of resisting death and closure. Unlike some of her Protestant predecessors and contemporaries . . . Meynell admired the independence of a language she saw as related to, but not entirely under the control of, its authors. . . . Meynell's confidence in language is intrinsic to her understanding of the relationality at the centre of life, and is underlined by a belief that words mediate rather than distort the divine, albeit in an incomplete fashion.
> (*Nineteenth-Century Religion and Literature*, 2006, 203)

Consequently, in many of her lyrics, she lays out a line of verbs, and other grammatical forms, too, as if she is wayfaring among them, exploring where they lead and what they reveal—still following, we might say, the *tropos* of Christ in creation, his attentive way of being and loving. Almost always, at least one verb (or adjective or noun) in a series is a turn of some kind, a startle or surprise that opens generously to something more—a reminder that the poet is not walking a well-trodden path, nor navigating with intent only to arrive, but rather wayfaring among words to see what turns up in the meshwork of language.

Two examples must suffice. In "Veneration of Images" (*Collected Poems*, 1913, 27), the poet describes a worshipping man "whose wide arms entreat,/Gather, clasp, welcome, bind,/Lack, or remember" (ll. 1–2). While the first five verbs (which include synonyms for embrace and lock) can readily be envisioned as the arm actions of a man standing before a holy image or statue, the last two verbs seem initially to be poetic missteps: how can arms lack (forget) or remember? Are those not the domains of mind or heart? But Meynell's word-path refuses body-mind-soul separation. Instead, her line of verbs highlights how the whole human entity is intricately *in-by-through-for*, such that we ought not to limit the potential meaning of outstretched arms to the habitual. Meynell pointedly follows her line of verbs with a habitual expression—"warm pulses beat/With love of thine own kind" (3–4)—as if to underscore the point, namely, that we already acknowledge physical-affective affiliation in some of our expressions; she is merely scaling up from pulses to arms, and from there, to the whole posture of the worshipper. For that is ultimately where the poem heads, following out its own cast lines. The unified body-soul of the observed worshipper, with his outstretched arms evoking the crucifix, prompts the poem's surprising, closing lines: "O flesh, O grief, thou too shalt have our knee/Thou rood of every day!" (7–8). The venerator becomes the venerated because in his "O flesh, O grief," he bears the image of all humanity. Thus, the poem's title, "Veneration of Images," walks away from its typical association into new territory as Meynell attends to where her verbs take her.

Similarly, in "To the Body," Meynell explores through a series of verbs the physical and spiritual topography of the human body as integrated with that of the rest of creation; though not part of her series, "close" again functions as a key verb. In this poem, the speaker apostrophizes the body as a royal and judicial court wherein sit its senses, sued by entities of earth and sky that desire relationship with them, a relationship the senses agree to enter; the resultant closing together of sense with world activates both entities—but only after the senses walk a full line of verbs to get from legal agreement to creative action. The arrival at "create" seems to startle even the poet, such that she holds the verb over the pause across two stanzas, repeats it, and follows it with a dash that catapults all that follows:

> Thou inmost, ultimate
> Council of judgment, palace of degrees,
> Where the high senses hold their spiritual state,
> Sued by earth's embassies,
> And sign, approve, accept, conceive, create;
>
> Create—thy senses close
> With the world's pleas. The random odours reach
> Their sweetness in the place of thy repose,
> Upon thy tongue the peach,
> And in thy nostrils breathes the breathing rose.
>
> To thee, secluded one,
> The dark vibrations of the sightless skies,
> The lovely inexplicit colours run;
> The light gropes for those eyes.
> O thou august! thou dost command the sun.
>
> Music, all dumb, hath trod
> Into thine ear her one effectual way;
> And fire and cold approach to gain thy nod
> Where thou call'st up the day,
> Where thou awaitest the appeal of God.
> (*Collected Poems*, 1913, 55)

Only when a sense-capacity of the human body closes with an element of the external world that is seeking that sense does sweetness or fragrance or colour come into being. Smell, taste, sight, hearing, touch, all become possible only when the body is fully part of the meshed world, its peaches, roses, skies, temperatures, all. It is the same *in-by-through-for* conception of the meshed world consistent across Meynell's poetry. But the poem's verbs take one more startling step: in the final stanza, the body-as-court calls up the day and awaits the appeal of God. One possible interpretation of this final line is that the body has made a decision it expects God to appeal, or that God might even be subordinate to the body's rulings, such that an appeal is necessary. As we noted with "Christ in the Universe," Meynell is not afraid to wander into surprising theological spaces, so such an interpretation is possible. But paired with "the world's pleas," "the appeal of God" seems more likely a step deeper into the idea of unity that the poem is pursuing. Just as the body has responded to the world's pleas for closing together, for embrace, for mutuality, so the body now waits for God's call for the same—at which point, it will respond in the same way, entering full relationship with God, without barrier of time and place, becoming finally its fullest self. Indeed, this is, ultimately, what the whole meshed world waits for. But while it waits, it also moves along lines that imitate God's own close, incarnated action with the world, similarly represented by Meynell through accumulated verbs in "Messina, 1908," where the poet addresses the "Lord, [who] hast hastened to retrieve, to heal/To feed, to bind, to clothe, to quench the brand,/To prop the ruin, to bless and to anneal" (*Collected Poems*, 1913, 42, ll. 7–9).

As noted earlier, Meynell does not restrict such abundance to verbs. At times, she roams with adjectives ("Christ in the Universe": "the secret, cherished, perilous,/The terrible, shamefast, frightened, whispered, sweet/Heart-shattering secret of His way with us," 10–12), and at times with nouns ("The Courts": "Their splendours, colours, avenues, arrays,/Their courts that run with wine," 3–4). But whatever the grammatical form, the effect is always one of exploring the multitude of lines cast out by an entity in the meshwork—a recognition that in the moral and spiritual topography of the world, a single word does not suffice to express the wonder and complexity of a thing. How can it, when—in Meynell's Catholic view—Christ wayfares in the entire meshwork of creation and not only among humans? In "A General Communion" (*Collected Poems*, 1913, 45), Meynell even describes the "devout people" (3) at mass as "a field of flowers" (9), a "thousand single central daisies" (9), each receiving "the whole of the devoted sun" (16)—a play on words that recalls "devoted Lord" (4), thus asserting that daisies and people alike are grown at unimaginable price (10–11). That unimaginable price and power Meynell is not afraid to explore by following pathways of thought that end in surprise, even shock. "The Crucifixion" (*Collected Poems*, 1913, 48) opens by exclaiming "man's capacity/For spiritual sorrow, corporal pain!" (ll. 1–2), then notes that Christ explored it the most deeply, but rather than die of it, instead "over the abyss/Of God's capacity for woe He stayed/One hesitating hour; what gulf was this?/Forsaken He went down, and was afraid" (ll. 13–16). "Veni Creator" (*Collected Poems*, 1913, 33) opens by asking God if he has left any path untrod for our sake, and answers yes before concluding, "Lord of Heaven,/Come to our ignorant hearts and be forgiven" (ll. 13–14). Christ, afraid? God, forgiven? These surely are erring feet in Catholic Christianity—or any Christianity. But for Meynell, they are not, because words for the poet are like steps for the wayfarer: they belong to the "child of process" for whom Christ's being both Way and Wayfarer is assurance that exploring possibilities—even possibilities from which one draws back—can be part of one's learning to inhabit fully the spiritual topography of the world, the meshwork of creation within which Christ walks.

Notes

1 Later in that same year, Orby Shipley included "Unto Us a Son Is Given" in his *Carmina Mariana, Second Series* (1902), 105–106, and noted his source as "From the 'Weekly Register,' 1897." However, I can find no trace of this publication or of a manuscript that

supports such a dating. On the possibility that Shipley might have misremembered his source, I therefore treat this poem as having first appeared in Meynell's 1902 collection, though I cite it from Meynell's 1913 *Collected Poems*, 32, where it appeared unrevised from 1902.

2 "The Fugitive" first appeared in *The Saturday Review*, no. 2675, Vol. 103, on 2 February 1907, p. 140 (the only poem in the number), before joining Meynell's 1913 *Collected Poems* (p. 46), unrevised except for how the epigraph is identified. I cite from *Collected Poems*.

3 "Christ in Portugal" was first published in the London *Tablet*, a weekly Catholic journal, Vol. 119, no. 3738, on 30 December 1911, p. 1045. In *Collected Poems* (p. 47), it is retitled as "In Portugal, 1912," possibly because additional fighting between anti-Catholic republicans and pro-Catholic monarchists in Chaves, Portugal, that year added further resonance to the poem. In this version, the final line of the poem, originally "The wine-press hides the unbidden Christ," is "The wine-press holds the unbidden Christ."

4 As Knight and Mason note, "The play of the word 'lent' on the period of abstinence that Christian tradition practises in the period immediately prior to Easter Sunday suggests that Meynell views the Incarnation as a more useful starting-point for theology than the focus on the Cross favoured by Protestantism" (*Nineteenth-Century Religion and Literature*, 2006, 202).

5 "Christ and the Universe" first appeared in *The Fortnightly Review*, Vol. 90 (New Series), no. 1038, on 2 October 1911, 712. Meynell published it next in *Collected Poems* (1913) as "Christ in the Universe," with eight other textual revisions. I cite from *Collected Poems* unless specified.

6 See Sharon Smulders ("Looking 'Past Wordsworth'," 2000) for more on this subject.

References

Anderson, Kathleen. "'I Make the Whole World Answer to My Art': Alice Meynell's Poetic Identity." *Victorian Poetry* 41, no. 2 (2003): 259–276.

Armstrong, Isobel. *Victorian Poetry: Poetry, Poetics, and Politics*. 2nd ed. New York: Routledge, 2019.

"close, v." *OED Online*. December 2022. Oxford: Oxford University Press.

Doane, George Washington. "Thou Art the Way." In *Hymns Ancient and Modern*. Compiled and arranged by William H. Monk. London: Novello, 1861. Hymn 162. Internet Archive 2013. https://ia903207.us.archive.org/16/items/modeance00chur/modeance00chur.pdf.

Fraser, Hilary. "Aesthetics, Visuality and Feelings in the Natural Theology of Gerard Manley Hopkins and Alice Meynell." In *Form and Feeling in Modern Literature: Essays in Honour of Barbara Hardy*, edited by William Baker with Isobel Armstrong, 88–99. London: Legenda Press, 2013.

Gray, Elizabeth F. "Alice Meynell, Poetry, and the Eucharist." *Christianity and Literature* 52, no. 2 (Winter 2003): 159–179.

Ingold, Tim. *The Life of Lines*. New York: Routledge, 2015.

Knight, Mark and Emma Mason. *Nineteenth-Century Religion and Literature: An Introduction*. Oxford: Oxford University Press, 2006.

Meynell, Alice. *Collected Poems of Alice Meynell*. New York: Charles Scribner's Sons, 1913. https://archive.org/details/collectedpoemsof00meynuoft.

———. *Later Poems*. London: John Lane, 1902. https://archive.org/details/laterpoems00meyniala.

———. *Other Poems*, 1896. Privately printed. Wesleyan University Library. https://onesearch.wesleyan.edu/discovery/fulldisplay/alma995066683903768/01CTW_WU:CTWWU. Accessed 18 February 2023.

———. *Poems*. London: Elkin Matthews and John Lane, 1893. https://archive.org/details/poemsmeynellalic00meynrich.

———. *The Poems of Alice Meynell*. London: Oxford University Press, 1940. https://archive.org/details/poemsofalicemeyn0000meyn.

Murphy, Patricia. *Reconceiving Nature: Ecofeminism in Late Victorian Women's Poetry*. Columbia: University of Missouri Press, 2019.

O'Connell, Rachel. "'Significant Negatives': Genre and Ethics in Alice Meynell's Familiar Essays, 1893–1909." *Cahiers victoriens et édouardiens* 74 (2011): 59–74.

Peterson, Linda. "Alice Meynell's 'Preludes', or Preludes to What Future Poetry?" *Victorian Literature and Culture* 34, no. 2 (2006): 405–426.

———. "Writing Nature at the Fin de Siècle: Grant Allen, Alice Meynell, and the Split Legacy of Gilbert White." *Victorian Review* 36, no. 2, Natural Environments (Fall 2010): 80–91.

Scholl, Lesa. *Hunger, Poetry and the Oxford Movement: The Tractarian Social Vision*. London: Bloomsbury, 2020.

Shipley, Orby. *Carmina Mariana, Second Series*. London: Burns & Oates, 1902. https://archive.org/details/carminamarianase00shipiala.

Smulders, Sharon. "Looking 'Past Wordsworth and the Rest': Pretexts for Revision in Alice Meynell's 'The Shepherdess'." *Victorian Poetry* 38, no. 1 (Spring 2000): 35–48.

Wirzba, Norman. "Creation *Through* Christ." In *Christ and the Created Order: Perspectives from Theology, Philosophy, and Science*, edited by Andrew B. Torrance and Thomas H. McCall, 35–53. Grand Rapids: Zondervan, 2018.

11 Flannery O'Connor's Integral Ecology

George Piggford, C.S.C.

The title of Pope Francis's 2015 encyclical *Laudato Sí* is taken from Francis of Assisi's "Canticle of the Creatures," composed c. 1224.[1] *Laudato Sí* means literally "Praise Be to You," and in his poem St. Francis praises the Lord "through our Sister Mother Earth, who sustains and governs us, and who produces various fruit with colored flowers and herbs" (*Francis of Assisi*, 1999, 113–114). For St. Francis, there is affinity between human beings and the natural world, just as for Pope Francis, "nature cannot be regarded as something separate from ourselves or as a mere setting in which we live." For the Pope, "we are part of nature, included in it and thus in constant interaction with it" (Francis, *Laudato Sí*, 2015, 105 [no. 139]). The Pope's encyclical emphasizes the symbiosis between human beings and the natural world. This interconnectedness, which the Pope terms "integral ecology," is rooted in long-standing Catholic approaches to creation. *Laudato Sí* has been powerfully informed by Ken Wilber's "integral theory," the liberation theologian Leonardo Boff, and the American ecotheologian Thomas Berry.[2]

In addition, Pope Francis's encyclical owes a significant debt to Pierre Teilhard de Chardin, the Jesuit mystic and paleontologist (1881–1955). Although Teilhard is mentioned only once in the Pope's lengthy letter, subtitled "On the Care of Our Common Home," its articulation of "universal communion" is modelled on the writing of the French priest, especially his "Mass on the World" (Francis, *Laudato Sí*, 2015, 65 [no. 83]).[3] Teilhard originally composed that meditation on the cosmic implications of the Catholic Eucharist in 1923, in China's Ordos desert.[4] "In the steppes of Asia," with "neither bread, nor wine, nor altar," Teilhard imagines a Holy Communion that is worldwide and ultimately universal, when "the contours of [God's] body melt away and become enlarged beyond all measure" and "the world is caught in the descending radiance of the heart of God" (*Hymn of the Universe*, 1960, 19, 34, 35).[5] In this mystical vision, the Earth and all life upon it become a consecrated Host held up to the cosmic God. Everything is holy, and in Teilhard's anagogical view, God's sacramental presence in nature is revealed, and all creation has become Christ. Teilhard's view heralds the "divinization" or "becoming divine" of the natural world and the entire cosmos.

Such a cosmic theophany might at first seem to be far afield from the work of American regionalist Flannery O'Connor (1925–1964), with the vast majority of her stories rooted in the distinctive red clay of her native Georgia. Like Teilhard,

DOI: 10.4324/9781003366744-11

however, O'Connor's view tended towards the mystical and the universal, especially in many of her stories' moments of grace. This feature of her writing is most famously present at the end of the short story "Revelation," first published in 1964. That story concludes with the shocked Ruby Turpin's view from her pig-parlour of an all-encompassing and upended human social order accompanied by "invisible cricket choruses" (O'Connor, *Complete Stories*, 1971, 509). The ultimate gesture of the story is a vision of the Communion of Saints representing racial diversity, varieties of ability, and all classes of society.[6] This "vast horde of souls" is "rumbling towards heaven" in a kind of cosmic communion line; all are "climbing upward" into the "starry field" of the visible universe. With its final word "hallelujah"—from the Hebrew root "praise be to you Lord"—the story provides a hopeful vision of the ultimate fate of both its protagonist Ruby and diverse humanity (508, 509). A vision that includes humans, nature, and the cosmos not only echoes Teilhard's "Mass" but also anticipates the current Pope's emphasis on universal communion—the cosmos itself understood as the Body of Christ.

Given O'Connor's astringent and sometimes brutal realism, however, such a magnanimous vision is a rarity. More typical is "A View of the Woods," a short story that calls its readers' attention to the divine and Eucharistic qualities in nature but examines more closely the failure of human beings to enact an ethic inspired by what Francis calls an "integral ecology," which is spiritual and physical, as well as cultural, economic, and ethical (*Laudato Sí*, 2015, 104 [no. 137]). Teilhard commented in his 1916 essay "Cosmic Life" that "there is a communion with God, and a communion with earth, and a communion with God through earth" (Teilhard de Chardin, *Writings in Time of War*, 1968, 14). What of those, however, who fail to perceive and act on this presence of God in the natural world? In "A View of the Woods" O'Connor emphasizes not only that divinity dwells within nature but also that our own planet and all life upon it is the wounded and bleeding Christ, especially in the face of industrialization and exploitation aided by technology. It is in suffering that the world and human beings find their place within Christ's "Mystical Body," in a process that Teilhard, following eastern Christian traditions, terms "divinization" (Teilhard de Chardin, *The Divine Milieu*, 1960, 7).[7]

Critics have emphasized the sacred qualities of the natural world in O'Connor at least since Louise Westling's important *Sacred Groves and Ravaged Gardens*, which deems O'Connor's "woods and pastures" to be "sacred precincts for the enactment of violent rituals intended to batter her characters into an awareness of their helplessness before God" (Westling, *Sacred Groves*, 1985, 156–157). This stress on the damage that humans inflict on nature, often aided by technology, is evident from the beginning of "A View of the Woods," which O'Connor completed by the end of 1956 (O'Connor, *Letters*, 1979, 175). In the opening scene of the story, we are introduced to a grandfather, Mark Fortune, and one of those grandchild doppelgängers familiar to us in O'Connor's work, in this case Mr. Fortune's granddaughter, Mary Fortune Pitts. They not only look alike physically but also, as the narrator asserts early on, "the spiritual distance between them" is "slight" (O'Connor, *Complete Stories*, 1971, 526). We meet them mutually staring at the toil of "two huge yellow bulldozers" moving soil in a "hole" that will eventually become the foundation of a fishing club at a new lake, modelled on Lake Sinclair, located just a few miles north of O'Connor's home of Andalusia and created by the State of Georgia and the Georgia Power Company in 1953.

The work of these bulldozers is not described in detail, except for a comment that they are "smooth[ing] out" the pit formed by a different contraption, called throughout the story simply "the machine" (O'Connor, *Complete Stories*, 1971, 528). Despite what many critics of the story have claimed, this machine is not itself a bulldozer but what is called an excavator, which is explicitly distinguished in "A View of the Woods" from its two attendant bulldozers and, later, a cement mixer.[8] The machine is most likely a hydraulic excavator mounted with a backhoe, which consists of a heavy jointed arm and a bucket at the end designed to draw earth back towards the machine (Haddock, *Giant Earthmovers*, 1998, 220). Such machines generally turn on a "swing" or swivel, which allows them to expel dirt at the edge of a deepening pit. They are propelled either on an endless caterpillar-like track or on vulcanized rubber wheels (Haddock, 1998, 200).

The narrative begins "the week before" the scene with the bulldozers—with a description of Mr. Fortune and Mary Fortune's habit of watching the excavator dig huge wounds into the blood-red Georgia clay:

> Mary Fortune and the old man had spent every morning watching the machine that lifted out dirt and threw it in a pile. . . . He sat on the bumper and Mary Fortune straddled the hood and they watched, sometimes for hours, while the machine systematically ate a square red hole in what had once been a cow pasture. . . . 'Any fool that would let a cow pasture interfere with progress is not on my books,' he had said to Mary Fortune several times from his seat on the bumper. . . . She sat on the hood, looking down into the red pit, watching the big disembodied gullet gorge itself on the clay, then, with the sound of a deep sustained nausea and a slow mechanical revulsion, turn and spit it up.
> (O'Connor, *Complete Stories*, 1971, 335)

The machine is consistently described in monstrous terms and with malevolent motives. It is in fact a mechanized version of Lucifer in Dante's *Inferno*, not masticating history's greatest traitors in the lowest pit of hell but chewing up and spitting out the earth itself.[9] The excavator eats but is never satisfied; it can chew but never swallow. It "systematically" damages the land not for its own sustenance but for the sake of what Mr. Fortune calls "progress."[10]

This version of progress represents, as Mark Graybill has noted, the "commercial and technological exploitation of nature," but it also anticipates many of the arguments in Pope Francis's environmental encyclical (Graybill, "Ecological Vision," 2011, 12).[11] "When people become self-centred and self-enclosed," Francis argues, "their greed increases. The emptier a person's heart is, the more he or she needs things to buy, own, and consume" (Francis, *Laudato Sí*, 2015, 148 [no. 204]).[12] For Francis, no amount of exploitative consumption can possibly nourish a human in the most meaningful sense. The repetitive scene of the excavator gluttonously "gorg[ing] itself on the clay" only to vomit it up immediately evokes the profound emptiness of this compulsion to consume.[13] There is clear evidence that for both the current Pope and Flannery O'Connor our global environmental crisis reflects and results from a deeper spiritual crisis and requires an "integral ecology," one that is concerned with cultural structures, economic justice, environmental awareness, and spiritual development (Francis, *Laudato Sí*, 2015, 104 [no. 137]). Indeed, a key insight in the encyclical is that social justice and ecological justice are integrally related. Francis thus calls for both a "bold cultural revolution" and a worldwide "conversion" to "a loving awareness that we are not disconnected from the rest of creatures, but joined in a splendid universal communion" (Francis, *Laudato Sí*, 2015, 88, 157, 158 [no. 114, no. 220]).[14] The Pope's language of "the rest of creatures" indicates a sense of communion that encompasses not only human animals but also all living entities.[15]

The vocabulary that Francis uses to describe his Catholic ecology is directly indebted to the writing of Teilhard de Chardin, especially "The Mass on the World." O'Connor did not live to see the publication of that work in English in 1965, but many of its ideas are present, although in a muted fashion, in Teilhard's *The Divine Milieu*, which O'Connor read and reviewed in early 1961 (O'Connor, *Presence of Grace*, 1983, 106). Given that my main literary focus here is "A View of the Woods," published in 1957, I am not as interested in the influence of Teilhard on O'Connor as I am in their similar environmental and Eucharistic visions and their anticipation of the approach that Pope Francis advocates in *Laudato Sí*. Mark Graybill, whose work is central to an understanding of what he calls O'Connor's "deep ecology," comments perceptively that O'Connor's fiction is "not so much . . . reflective of Teilhard's thought," as it is "complementary to it" (Graybill, "Ecological Vision," 2011, 6).[16] That is, even before she read his work, O'Connor in her fiction and other writing foresees Teilhard's cosmic vision.[17] When O'Connor finally turned to the controversial Teilhard in the early 1960s after a conversation with her editor Robert Giroux, what she found was a kindred intelligence and similar capacity for mystic vision.[18] This congruence is evident in a number of her letters and is perhaps best summed up in her review of *The Divine Milieu*. There, O'Connor declares: "It is doubtful if any Christian of this century can be fully aware of his religion until he has reseen it in the cosmic light which Teilhard has cast upon it" (O'Connor, *The Presence of Grace*, 1983, 108). Teilhard's efforts to reconcile Christian belief and Catholic sacramentalism with both biological and cosmological evolution offered O'Connor fresh insight into the relationship between God and creation.

Teilhard's claims about the unifying and trans-historical qualities of the Catholic Eucharist are evident in the only passage that O'Connor marked in her own copy of *The Divine Milieu*, currently in the archive at Georgia College and State University in Milledgeville (Kinney, *Library*, 1985, 33–34). In it, Teilhard notes that every Mass includes a priest and words of consecration, which in his time were universally *hoc est corpus meum* ("this is my body"). Each Mass, however, points to something even larger. As Teilhard expresses it, "the great sacramental operation does not cease at that local and momentary event." He continues:

> Even children are taught that, throughout the life of each human being and the life of the Church and the history of the world, there is only one Mass

> and one Communion. Christ died once in agony. Peter and Paul receive communion on such and such a day at a particular hour. But these different acts are only the diversely central points in which the continuity of a unique act is split up and fixed, in space and time, for our experience.
>
> (Teilhard de Chardin, *Divine Milieu*, 1960, 102)

In every Mass, the transcendent God intersects with chronological, human time, and the infinitely vast becomes present in the tangible, physical forms of bread and wine.

Further, for Teilhard the Incarnation is "realized, in each individual, through the Eucharist." Every human being who approaches God through a Eucharistic meal becomes intimately close to Christ, a partaker of the divine nature. This sacramental experience is, in fact, universal, communal, and trans-historical: "All the communions of a life-time are one communion. All the communions of all men [and women] now living are one communion. All the communions of all, present, past and future, are one communion" (Teilhard de Chardin, *Divine Milieu*, 1960, 102). The Eucharistic meal, in other words, always takes its participants beyond a specific sacramental action towards divinization, what Teilhard also calls "christification," that is ultimately cosmic (101).[19] Daniel Scheid, in *The Cosmic Common Good*, builds on this central point when he declares, "Communion inheres into the makeup of the universe itself. It springs from and reinforces the intrinsic goodness of every individual species and creature" (Scheid, *Cosmic Common Good*, 2016, 75).[20] Mass connects the individual not only to the historical person of Christ—God in the flesh from a Catholic perspective—but also to the universal Christ. Mass makes, or ought to make, its human participants more profoundly aware of the communion inherent in nature and in the universe, and the sacrament performs a central role in the transformation bringing all of creation into the Body of Christ.[21]

Teilhard calls the universal Christ the "Omega Point," which is, to borrow Ilia Delio's phrase, "the power of evolution . . . who is within and ahead" (Teilhard de Chardin, *Phenomenon of Man*, 1968, 257; Delio, *Making All Things New*, 2015, 93). Christ as Omega is the motivating force of the universe, the proper end of all human action, and the inevitable culmination of all processes of generation, both cosmological and biological.[22] It is this very idea that provided O'Connor with the title of her second collection of short stories, *Everything That Rises Must Converge*, a phrase inspired by Teilhard's vision in *The Phenomenon of Man* of evolution as a pyramidal cone, "whose apex is supported from below" (Teilhard de Chardin, *Phenomenon*, 1968, 270). Kathleen Duffy points out in *Teilhard's Mysticism* that the act of creation is ongoing: "the world is continually being created, allured toward higher levels of convergence by the personalizing and unifying presence of Omega, the Cosmic Christ" (Duffy, *Teilhard's Mysticism*, 2014, 118). In this formulation, Christian divinization was initiated at the events surrounding the crucifixion of Jesus of Nazareth and it continues, in fact evolves, at every Eucharistic celebration. Christ is, in Duffy's rendering, the "inner face," the personal and spiritual quality, of the evolutionary principle itself. The beginning of the cosmos anticipates the emergence of life on Earth and its natural environment, which provided the setting for God's incarnation as a human being. As stewards of our cosmic and global setting, we continue the process of christification in an effort that Teilhard calls "noogenesis," which refers to the evolution of the human mind into an ever more ethical (and Christ-like) relationship with the world.[23] Catholic Mass is here understood as facilitating this development of human awareness and the ongoing transformation of all matter and energy into the divine life.

As an orthodox Roman Catholic, it was for O'Connor of utmost importance that at every validly celebrated Catholic Mass, bread and wine are transubstantiated into the Body of Blood of Jesus Christ. One of her best-known quotations emphasizes this belief. In a 1955 letter to her friend Elizabeth Hester, O'Connor describes a dinner party in New York City that included a conversation about the Eucharist, "which I, being the Catholic, was obviously supposed to defend." When other guests proffered a symbolic interpretation of the sacrament, O'Connor replied, "Well, if it's a symbol, to hell with it" (O'Connor, *Letters*, 1979, 125). This insistence on physicality and the body of an incarnate divinity in Catholic Communion has unsettled many throughout history. The emphasis on the bread and wine, products of field and vine and of human labour, as truly Christ's Body and Blood is long-standing in the Catholic tradition. By the nineteenth and twentieth centuries, Catholic theology highlighted more than any other dogma the reality and primacy of the Eucharistic Sacrament as an outward sign not only of God's grace but of God's Real Presence and mysterious indwelling in the world.[24] Timothy Vande Brake has pointed out that there are environmental consequences to such an idea of the Eucharist: this Teilhardian and cosmic "view posits that God's real presence in the humble bread and wine of the Eucharist is God's continual affirmation that the world of things is very good," to be treated with utmost care and respect (Vande Brake, "Thinking Like a Tree," 2011, 30).

It is at this point that a question arises, one pondered by both Teilhard and O'Connor, despite the latter's emphasis on Eucharistic transubstantiation: Why do only bread and wine become Christ? If these items might be transubstantiated, then why not everything? Why does divinization cease with processed wheat and fermented grapes? In 1923, at his archaeological site in the Ordos desert, the priest Teilhard woke up on a Sunday morning with no bread and no wine at hand. What he substituted in their place is recorded in "The Mass on the World."[25] Here, his Eucharistic theology is much more radical than what he expresses in *The Divine Milieu*, and in fact encompasses all the natural world, living and dying. In this mystical prayer, the Eucharist is freed not only from the walls of churches and the rites of the Mass but also from the specificity of the elements of bread and wine:

> Over every living thing which is to spring up, to grow, to flower, to ripen during this day say again the words: This is my Body. And over every death-force which waits in readiness to corrode, to wither, to cut down, speak again your commanding words which express the supreme mystery of faith: This is my Blood.
>
> (Teilhard de Chardin, *Hymn of the Universe*, 1960, 23)[26]

Here, Earth herself becomes the altar of sacrifice, and the Host is associated with the life-force of the natural world. It is, indeed, life or really rebirth—"to spring up, to grow, to flower, to ripen"—that transubstantiates the world, that makes it into the Body of Christ. The element of wine is associated with the "death-force" that becomes transubstantiated into Christ's Blood. With the sheen of nature, living and dying, on its surface, the world becomes a planet-sized Host lifted up to the God of the cosmos.[27]

Teilhard's approach to the Eucharist was too extreme for the Catholic authorities of the mid-twentieth century, and a *monitum* or "warning" was issued against his writing in 1962.[28] From a Catholic perspective, one danger of this approach is pantheism, a religious view that O'Connor emphatically rejects in a 1956 letter to Elizabeth Hester: "Remember that I am not a pantheist and do not think of the creation as God, but as made and sustained by God" (O'Connor, *Letters*, 1979, 126). Because a pantheistic conflation of creation with Divinity undermined his model of evolutionary spirituality, such a position was also disavowed by Teilhard, most clearly in *The Divine Milieu.* In that book, he insists that "the sojourner in the divine milieu is not a pantheist." This is so because pantheism offers "only fusion and unconsciousness" in its vision of "perfect universal union" (Teilhard de Chardin, *Divine Milieu*, 1960, 93). What Teilhard offers is, instead, a Christian and Catholic paradox of individuality that finds its distinctiveness in union with God: "*to be united* (that is, to become the other) *while remaining oneself*" (94).[29] Therefore, Christians may "plunge" themselves completely into the divine milieu—the natural world—"without the risk" of becoming a pantheistic monist, someone who perceives God and the cosmos to be one and the same (94).[30] Nevertheless, as his contemporary and fellow Jesuit, Henri de Lubac notes, Teilhard's mysticism is sometimes misread in this fashion. Teilhard for de Lubac remains "the most outspoken opponent of pantheistic concepts of the Godhead" because he articulates so clearly the simultaneous transcendence and profound immanence of a God who is both "further than everything and deeper than everything" (de Lubac, *Faith*, 1965, 22, 24). This understanding of God's presence in the natural world comports well with (1) the vision of the Jesuit order's founder, Ignatius of Loyola; with (2) de Lubac's theological interpretation of Teilhard; and with (3) the poetry of yet another Jesuit, Gerard Manley Hopkins. This last writer was especially influential upon O'Connor's mystical ecopoetics.

In *Hymn of the Universe* Teilhard admits that "like the monist I plunge into the all-inclusive One," but here the crucial distinction between "like" and "as" is most relevant. To be *like* a monist is to differentiate oneself through comparison. The Teilhardian mystic loses the human "ego" in the divine One only in order to "find in it the ultimate perfection of . . . individuality" (Teilhard de Chardin, *Hymn*, 1960, 26). By becoming Christ, in other words, we become most ourselves. This paradoxical formulation provides a striking echo of Hopkins's famous sonnet "As Kingfishers Catch Fire," in which the human person "acts in God's eye what in God's eye he is—Christ." Christ and the self are not (yet) one but are in a playful and powerful relationship, with the self undergoing a process of Christification: "Christ plays in ten thousand places, lovely in limbs and lovely in eyes not his" (Hopkins, *Major Works*, 2002, 129).[31] Likewise, the transubstantiated species of bread and wine retain their distinctive sensate qualities in orthodox Catholic theology. They become Christ while remaining in texture, taste, and materiality what they have always been. According to the *Baltimore Catechism*, which was the manual of faith for American Catholics of O'Connor's generation, these items become Christ's Body. However, "the color, the taste, and whatever appears to the senses" remain (Third Plenary Council 2010, 197 [Q. 885]). The Eucharistic elements are utterly transformed while remaining their distinct species, and according to Teilhard the same is true of the human species understood as becoming the Body of Christ. Christ seeks to dwell everywhere in creation in a mode that is not possessive but perfecting. As Pope John Paul II has memorably phrased it, "We are asked to become one. We are not asked to become each other" (John Paul II, *Letter*, 1988). Humans are to be found in nature, not lost in it; integrated, not obliterated.

Recent popes have come around to a view of the Eucharist comparable to Teilhard's, where the cosmos itself is envisioned as becoming Christ. John Paul II's encyclical *Ecclesia et Eucharistia* declares that "the Eucharist is always in some way celebrated *on the altar of the world.* It unites heaven and earth. It embraces and permeates all creation."[32] This language evokes Teilhard's notion of convergence at the Omega Point. Of all Catholic leaders, Pope Emeritus Benedict XVI has made the greatest effort to rehabilitate Teilhard's reputation and ideas. A 2009 homily features a repetitive emphasis on the main theme of "The Mass on the World":

> The role of the priesthood is to consecrate the world so that it may become a living host, a liturgy: so that the liturgy may not be something alongside the reality of the world, but that the world itself shall become a living host, a liturgy. This is also the great vision of Teilhard de Chardin: in the end we shall achieve a true cosmic liturgy, where the cosmos becomes a living host.
>
> (Benedict XVI, *Homily*, 2009)

At the end of cosmic history, all creation will be transubstantiated, God will dwell in all as all. Only then will God complete the great poem, the Mass or the Eucharistic liturgy. This process began with the creation of the cosmos itself, continued with the mystery of divine Incarnation, and moves forward even now. Pope Francis, echoing Teilhard, calls the Eucharist "an act of cosmic love" (Francis, *Laudato Sí*, 2015, 168 [no. 236]). In his view, we are to care for our common home, which includes not only the natural setting of Earth but of the cosmos itself. We are to treat all reality as if it were, at least potentially, Christ.

While it is true that Flannery O'Connor did not know "The Mass on the World," she did imagine at least the possibility of something similar, and the best evidence for this can be found in her most ecological of stories, "A View of the Woods." In that story, we find in Mr. Fortune a potential though ultimately failed mystic in the mold of Teilhard. This character expresses, in Christine Flanagan's phrase, "a malevolent desire for domination" and identifies with the "monstrous, gluttonous excavator" ("From Earth to Eternity," 2016, 165). At the same time, Mr. Fortune's vision is frequently haunted by the Christ-like and even Eucharistic qualities of the pastures and woods that he is serially selling off and enabling the destruction of, all in the name of "progress." In the story, Mr. Fortune is provided an opportunity to become aware of the divinity of the natural world before him and indeed of its Christ-like qualities. He is invited into an attitude of "accountability to the environment," as Lawrence Buell puts it (*Environmental Imagination*, 1996, 7), but he turns away from the suffering Christ hidden in the woods in the name of an ultimately self-destructive "progress." His failed eco-mysticism, his inability to perceive the divine indwelling in nature, will prove lethal both to Mr. Fortune and to his granddaughter Mary Fortune Pitts.

"A View of the Woods" begins and ends with a focus on the malevolent machine, the Earth-wounding excavator that serves as the agent of Mr. Fortune's will to progress, his primary instrument for developing the heretofore relatively undisturbed landscape of central Georgia. This project might helpfully be situated within what historian C. Vann Woodward memorably calls the "bulldozer revolution," a phrase meant to evoke the massive changes to the topography of the postwar South owing to an economic boom and the belated industrialization of this region of the United States (Woodward, *Burden*, 2008, 6). A concurrent process was suburbanization, and the creating of artificial lakes such as Georgia's Lake Sinclair raised property values for newly lakefront parcels. These alterations to waterways and the land gave rise to infrastructure like the "paved highway" that Mr. Fortune imagines supplanting the "clay road" currently running in front of his home (O'Connor, *Complete Stories*, 1971, 337). Once the road has been paved, Fortune "wanted to see a supermarket store across the road from him, he wanted to see a gas station, a motel, a drive-in picture-show within easy distance" (337). Such structures would, in his imagination, comprise a new town: Fortune, Georgia. This legal renaming represents his desire not only to alter the environment around him but also to rechristen it as his namesake and thereby claim permanent possession of his property.

Mark Fortune's favourite grandchild, Mary, begins the story as a proponent of her grandfather's notion of "progress" and is also his double. As both of them sit on a car hood and watch the "disembodied gullet" of the excavator continue to "gorge" on the Georgian clay, the girl's face is described as "a small replica of the old man's" that "never lost its look of absorption" in the Earth-wounding work of the machine (336).[33] When, however, Mr. Fortune makes the decision to sell "the lawn" or open field in front of the house where Mr. Fortune and Mary live with the rest of her family, the Pittses, the granddaughter rebels against him (342). She is concerned that the gas station proposed for this parcel will take away the family playground and the area where their calves graze. Most importantly, it will obscure their view of the woods. What Mr. Fortune cannot or will not realize is the value of this undeveloped land across the road from "the lawn" (342). His attitude towards property is economic and exploitative. In his view, land is valuable only for its potential to be developed and to bring to reality his destructive idea of a town named Fortune. Nevertheless, the woods have something to try to teach him about the importance of the natural world and of its Christ-like and Eucharistic qualities if only he will grant them his sustained attention.

Not long after his argument with Mary, Mark Fortune rises after his afternoon nap and gazes at the woodlands across the lawn, but "[e]very time he saw the same thing: woods—not a mountain, not a waterfall, not any kind of planted bush or flower, just woods" (348). Rather than perceiving the diversity of the natural scene, it appears to him as an undifferentiated whole, simply "woods." But in the afternoon light "woven through them," each "pine trunk stood out in all its nakedness" (348). The sense of this description is that there is something vulnerable about each tree, something stripped bare, and, as we will see, something Christ-like. Nevertheless, Mr. Fortune seems incapable of peering deeply into this scene and recognizing the woods as something other than property to be owned and sold. In the language of Pope Francis, Mr. Fortune lacks an "integral ecology" that might help him to understand his interconnectedness not only to other human beings, especially those suffering and in need, but also to the trees and the complex ecosystem into which human and nonhuman entities are "woven."[34]

Mr. Fortune rises out of bed a second time to gaze at the woods, but he is only "reconvinced of his wisdom in selling the lot" (348). A powerful moment of mystic vision occurs, however, upon Fortune's third view of the woods. At this point, he begins to develop what Scott Slovic discerns in the writing of Henry David Thoreau, a "habit of attention" (see Slovic, *Seeking Awareness*, 1992, 4–7):[35]

> The third time he got up to look at the woods, it was almost six o'clock and the gaunt trunks appeared to be raised in a pool of red light that gushed from the almost hidden sun setting behind them. The old man stared for some time, as if for a prolonged instant he were caught up out of the rattle of everything that led to the future and were held there in the midst of an uncomfortable mystery that he had not apprehended before.
>
> (O'Connor, *Complete Stories*, 1971, 348)[36]

This blood-like "gush" of red light leads him to a more specifically Christian insight into the mysterious nature of the woods. He perceives the mystery "as if someone were wounded behind the woods and the trees were bathed in blood" (348). The sense here is that the woods are suffering from the damage being inflicted on the nearby sections of land that Fortune has already begun to develop. Through the mysterious interconnectedness of the natural scene, these "naked" trees suffer from the damage being done elsewhere on Mr. Fortune's property. There is a hint of futurity here, thus a transcendence of diachronic temporality: "The road would soon be paved." These woods too will soon be cleared, and the "red trunks" rise up in the "black wood" in an effort to communicate that they are sacred precincts, that they are holy—that the world, in the words of Pope John Paul II, is an altar, and they are the sacrifice (348).

To say that Christ is present in the woods and the trees comes as no shock to any traditional reader of "A View of the Woods." The woods, after all, "walk" like Christ "across the water" on the very first page of the story, and O'Connor herself affirmed that only "the woods" in the story are "pure enough to be a Christ symbol if anything is" (O'Connor, *Complete Stories*, 1971, 335; *Letters*, 1979, 190). As the trees are felled and the blood-red Earth is gouged out by the excavator, Mr. Fortune, despite his spiritual impoverishment, begins to perceive someone wounded and naked within and behind them. The woods are calling the stubborn and greedy

Mr. Fortune to "a sense of deep communion with the rest of nature," which "cannot be real if our hearts lack compassion and concern for our fellow human beings" (Francis, *Laudato Sí*, 2015, 70 [no. 91]). Such lack of compassion characterizes Mr. Fortune as he rejects a sense of integration with nature and thereby resists his own divinization. He is, if you will, the excavator gorging himself on the clay of his own property, eating and not being satisfied.

Near the end of the story, a violent encounter between grandfather and granddaughter occurs, as they are unable to agree on the fate of the woods. After Mr. Fortune finalizes the sale of the property, he is determined to whip Mary in punishment for her stubbornness and disobedience. As she fights back, she takes on qualities of the natural world itself: her "small rocklike fists" batter Mr. Fortune while "five claws"—that is, her fingers—dig into "the flesh of his upper arm" (O'Connor, *Complete Stories*, 1971, 354, 355).[37] Eventually, however, Fortune is able to get the better of the child. His behaviour mimics the excavator's: he grips his hands "tight around her neck," lifts her head into the air, and beats Mary's head "hard against the rock that happened to be under it" (355). He does this again and again. At this point, Mr. Fortune remembers that Mary is his "image," although now "conquered" by his act of hollowing out her skull (355).

After murdering his granddaughter for her opposition to his project of developing the "lawn" and blocking a view of the woods, Fortune experiences a heart attack and is felled like so many of the pines on his own property. From his vantage point of lying prone on the earth, he looks up "helplessly along the bare trunks into the tops of the pines" (356). Meanwhile, his heart enlarges or expands, and he imagines an escape across the artificial lake and into the distant woods of his earlier vision. The mysterious and divine qualities of the natural scene continue to beckon him. In his dying imagination, however, the "gaunt trees" close ranks and march "across the water and away into the distance" (356). Similar to the beginning of the story, the woods walk across the (blood) "red corrugated lake" away from Mr. Fortune. Lying upon the red Georgian clay near his dead granddaughter, his only other company is the "huge yellow" excavator "gorging itself," as ever, "on clay" (335, 356). There is a sense of blood sacrifice in this ending, an offering up of not only a Eucharistic sacrifice but also a human one. The work of the machine reaches its apex with violent death, which is perhaps the inevitable end of efforts to exploit and consume, to dam rivers, to replace lawns with gas stations and woods with fishing clubs. Such activities taken to their extreme destroy not only the ecological balance but also those we love and finally, as in Mr. Fortune's case, our very selves.[38]

In her essay "The Nature and Aim of Fiction," likely written in early 1957, Flannery O'Connor invokes the habit of attention so important in the Thoreauvian tradition of American nature writing in order to avoid such self-destructive, even apocalyptic, attitudes towards the world. She also advocates for a sense of universal communion: "The longer you look at one object, the more of the world you see in it, and it's well to remember that the serious fiction writer always writes about the whole world, no matter how limited his particular scene" (O'Connor, *Mystery and Manners*, 1969, 77). This attentiveness reveals for O'Connor, as it reveals at least momentarily for the reluctant Mr. Fortune, that "Christ plays in ten thousand places" and that to wound Earth is to do harm to the Body of Christ. When O'Connor makes the point that "the bomb that was dropped on Hiroshima affects life on the Oconee River," she means this both physically and spiritually (O'Connor, *Mystery and Manners*, 1969, 77). It cannot be a coincidence that the Oconee is the river that was dammed to produce Lake Sinclair, fictionalized in "A View of the Woods." Everything is interconnected and thus affects every other element in the created order, and certainly the massive damage wreaked on the natural environment by the dropping of the atomic bomb on Hiroshima on August 6, 1945, reverberated across the world and through time. O'Connor composed "A View of the Woods" 11 years after that event, and her blood-red Earth and suffering, Christ-like trees continue to bear the wounds of that atrocity and its own fearsome instrument of death.

If there is any hope at the end of O'Connor's ecological tale, it emerges ironically from the deaths of Mark Fortune and his granddaughter. Her grandfather was preparing Mary to carry on his work of "progress" after his death, but that legacy is arrested first by her rebellion, then by the old man's murder of his heir apparent. Their corpses thus become potential compost for the rejuvenation of the woods. Although Mr. Fortune drafted a secret will leaving his property in trust to Mary, it is her family, the Pittses, who are the most likely heirs at the story's conclusion. O'Connor's original coda indicates this: "Pitts [Mary's father], by accident, found them that evening. He was walking home through the woods about sunset. The rain had stopped but the polished trees were hung with clear drops of water that turned red where the sun touched them" (O'Connor, *Letters*, 1979, 190).[39] Here Teilhard's "life force," in the form of pure water, is combined with the "death force," as the drops turn a bloody red. It is surely no coincidence that this combination echoes the holy water mixed with Communion wine that is prepared for Consecration at every Catholic Mass.

Although O'Connor wrote to Elizabeth Hester that "I don't really think" this draft ending "does anything for" the story (O'Connor, *Letters*, 1979, 190), the image combining life and death reaffirms the Eucharistic connotations of "A View of the Woods" and situates it within the broader Teilhardian vision of a "Mass on the World." For Teilhard, "everything around me is the body and blood of the Word," and he asks, echoing Christ in the Garden of Gethsemane, "How could I refuse this chalice"? This Communion is complete when "not only the life-bringing forces, but also those which bring death" converge in God's "special" and "vitaliz[ing]" love (Teilhard de Chardin, *Hymn of the Universe*, 1960, 28, 31). No doubt such mystical language would provide only cold comfort to the story's deceased protagonists, or really antagonists, and in any case the published version of O'Connor's story ends with a pessimistic focus on the monstrous excavating machine. "A View of the Woods" finally provides its readers with only a very muted notion that the survivors who appreciate the woods might in the end inherit and perhaps find some way to preserve them.[40]

"A View of the Woods" poses a basic question: whose legacy will endure—Fortune or Pitts? To use the language of the present volume: what will it be—self-destruction or a renewed sense of the divine indwelling within the natural world? In the midst of their disagreement over selling the woods, the grandfather asks the girl, "Are you a Fortune . . . or are you a Pitts? Make up your mind" (O'Connor, *Complete Stories*, 1971, 351). When Mark Fortune and his granddaughter are engaged in their final argument, the old man insists that Mary accept her identity as a Fortune and disavow her immediate family, the Pitts. He wants her to surrender to his attitudes and to the idea that she, like her grandfather, is destined to become the namesake of Fortune, Georgia. Mary strenuously resists, saying emphatically, "I'm PURE Pitts." The rebellion against her grandfather's attitudes will lead to her violent death. Through this familial struggle, the story asks: will an attitude of possession and ultimately unsatisfying consumption continue to wound Earth in the name of development and "progress"? Or will a more integrated and Eucharistic "view of the woods" raise awareness about ongoing damage to God's creation, generate a renewed sense of wonder and praise, and lead to ameliorative action?

As Pope Francis has persuasively explained, establishing an integral ecology is no mean task. Many spiritual traditions, including the Roman Catholic, have sought for centuries to raise awareness of the divine indwelling in the created order and the entire cosmos. This is one of the legacies of Pope Francis's namesake St. Francis of Assisi. At the same time, a perception of human integration with nature remains a profound challenge owing to attitudes of consumption championed by many Christians, along with the monstrous machines that have long seemed ascendant in the so-called "developed" world. Nevertheless, O'Connor, writing as a devout Catholic, is emphatic that wanton harm to any of God's creation is ultimately self-destructive, both individually and for the human species. For O'Connor and for Pope Francis, as for Teilhard de Chardin, we are called not to exploitation but to the cultivation of habits of attention, an ethic of care for the natural world, and participation in universal communion. In such ways we welcome, rather than futilely resist, the divinization of the cosmos.

Notes

1 A version of this essay is appearing as "Laudato Sí, 'The Mass on the World,' and Flannery O'Connor's Eucharistic Ecology," published in the Vernon Press volume *Weaving Words into Worlds*, edited by Caroline Durand-Rous and Margot Lauwers. Reproduced here by kind permission of Vernon Press.

2 Ryszard Sadowski ("Inspirations," 2016, 76–81) has persuasively argued that, in addition to Francis of Assisi, Pope Francis's *Laudato Sí* is profoundly indebted to these three pioneers of integral ecology.

3 On universal communion, see Francis, *Laudato Sí*, 2015, 69–72 (no. 89, no. 92).

4 Although Teilhard did not compose "The Mass on the World" until 1923, he had considered the theme of "perpetual communion" as early as 1906 while on retreat at Sidi Gaber in Egypt. See King, *Teilhard's Mass*, 2005, 2.

5 This vision is echoed in *Laudato Sí*: "The ultimate destiny of the universe is in the fullness of God, which has already been attained by the risen Christ, the measure of the maturity of all things" (Francis, 2015, 65 [no. 83]).

6 Those in Ruby's vision include (1) whites and "black[s] . . . in white robes"; (2) the physically and mentally disabled, described as "freaks and lunatics"; and (3) different economic classes, from the lower, pejoratively described as "white trash," to the upper, namely the "tribe" of the "respectable" and financially well-off (O'Connor, *Complete Stories*, 1971, 508).

7 For an insightful analysis of this idea in Teilhard in relation to eastern Christianity, Maximus the Confessor, and Thomas Merton, see St. John, *Contemplation*, 2011, 1–26.

8 On the nature, operation, and Dantean significance of the central excavating machine in "A View of the Woods," see Piggford, "Excavator," 2018.

9 Pertile, "Introduction," 2007, 89 points out that Dante's Lucifer "is a machine totally dehumanized and desensitized, that is to say, totally deprived of either emotions or intelligence." O'Connor studied *The Divine Comedy* as a graduate student at the University of Iowa and possessed a well-read copy of the Carlyle-Okey-Wicksteed translation; see Kinney, *Library*, 1985, 134–135.

10 Davis, "Grace in the Machine," 2009, 21 perceptively notes that O'Connor's stories often "privilege the artifacts of science and industry by making these very things the new bearers of the sacred," as for example the telescope in "The Lame Shall Enter First," 1962. In "View," however, it is the woods themselves that are portrayed as sacred and the excavator as a desecrator of this sacred space.

11 Thomas Berry in *The Great Work* describes this idea of "progress" in industrialized North America as a "psychic compulsion" that "drove the commercial and industrial entrepreneurs as well as the scientists and engineers in their work." Such compulsion resulted from a transition from an agricultural, "organic economy" to an "extractive economy" (Berry, *Great Work*, 1999, 140, 138).

12 A character similar to Mr. Fortune in "A View of the Woods" is Mrs. May in O'Connor's "Greenleaf," 1956. When Mrs. May gazes upon "green pastures" and "a black wall of trees" through a window in her home, she surveys property that she possesses and that might bring her profit. More profoundly, she perceives nature only in relation to herself: "When she looked out of any window in her house, she saw the reflection of her own character" (O'Connor, *Complete Stories*, 1971, 321).

13 This scene foreshadows Young Tarwater's unhealthy relationship with food in O'Connor's 1960 novel *The Violent Bear It Away*. His true hunger is for Christ, the Bread of Life, so "city food only weakened him" (O'Connor, *The Violent*, 1960, 161). On the Eucharistic qualities this novel, see Srigley, "Asceticism," 2012, 201–210.

14 This emphasis on ecological conversion builds on comments by John Paul II, Benedict XVI, and the Ecumenical Patriarch Bartholomew. A 1990 speech of John Paul II calls for a new relation between humanity and the earth: "Faced with the widespread destruction of the environment, people everywhere are coming to understand that we cannot continue to use the goods of the earth as we have in the past" (John Paul II, "Message," 1990, no. 1). In a 2007 address Benedict XVI unambiguously connects social justice to ecology: "The worsening scandal of hunger . . . reminds us of the urgent need to eliminate the structural causes of global economic dysfunction and to correct models of growth that seem incapable of guaranteeing respect for the environment and for integral human development" (Benedict XVI, "Address," 2007). On the relevance of Bartholomew's environmentalism to Francis's integral ecology, see Francis, *Laudato Sí*, 2015, 13–15 (nos. 7–9).

15 It is important to note here that Francis's vision of universal communion does not explicitly extend to nonliving entities, in contrast to Teilhard de Chardin and, later, Thomas Berry. For Berry "Every being enters into communion with other beings," and such beings include "Trees . . . insects . . . rivers . . . [and] mountains," entities, that is, that are biological, chemical, and geological (Berry, *Great Work*, 1999, 4, 5).

16 For Graybill "deep ecology" emphasizes "biocentric equality"; that is, the value of nonhuman life is not dependent on its relative usefulness for human beings ("Deep Ecological Vision," 2011, 2). This ecological tradition is most closely associated with the Norwegian philosopher Arne Naess and with Bill Devall and George Sessions' *Deep Ecology* (1985).

17 See, for example, O'Connor's 1952 novel *Wise Blood* and its description of the night sky over the fictional city of Taulkinham: "The black sky was underpinned with long silver streaks that looked like scaffolding and depth on depth behind it were thousands of stars that all seemed to be moving very slowly as if they were about some vast construction

work that involved the whole order of the universe and would take all time to complete" (33). Here one finds a vision of a cosmos under perpetual construction, persistently, even consciously, evolving.

18 See Gooch, *Flannery*, 2009, 322–323. On the controversial nature of Teilhard's writing from the vantage of the Catholic magisterium, see later discussion.

19 Ilia Delio clarifies this term in an ecological context: "we are to 'christify' the world" (Delio, *Making All Things New*, 2015, 94); that is, human beings ought to become Christ-like in our own dealings with the earth and to become aware that Christ's conscious presence pervades the cosmos.

20 Scheid borrows his notion of communion from Thomas Berry and applies it to the traditions of Catholic social thought and to ecology.

21 Michael P. Murphy has most helpfully connected Teilhard's idea of cosmic liturgy to O'Connor's mystical realism: "human fallenness is divinely encompassed and consecrated in her fiction as if in a liturgical space—a literary sacramentality, like liturgy itself, which is both precise in its care and cosmic in its scope" (Murphy, "Breaking Bodies," 2017, 53).

22 For an illuminating exploration of the connotations of Teilhard's term "omega," which signifies "a unity both within and without evolution," see Grim, "President's Corner," 2016.

23 In Teilhard's parlance, "noogenesis" refers to the process of evolving the "noosphere," "a human sphere of consciousness around and over the Earth" (Grim, 2016, 1). According to Delio this process is "psycho-social," and she characterizes the noosphere as "a planetary neo-envelope *essentially linked to the biosphere* in which it has its root, yet is distinguished from it. It is the natural culmination of biological evolution and not a termination of it" (Delio, *Making All Things New*, 2015, 109). For an exploration of the influence of Teilhard's notion of noogenesis on O'Connor, see Watkins, *Flannery O'Connor*, 2009, 39.

24 See, for example, Paul VI's encyclical *Mysterium Fidei:* "No one can fail to see that the divine Eucharist bestows an incomparable dignity upon the Christian people. For it is not just while the Sacrifice is being offered and the Sacrament is being confected, but also after the Sacrifice has been offered and the Sacrament confected—while the Eucharist is reserved in churches or oratories—that Christ is truly Emmanuel, which means 'God with us.' For He is in the midst of us day and night; He dwells in us with the fullness of grace and truth" (Paul VI, *Mysterium Fidei*, 1965, no. 67).

25 Thomas M. King stresses, "The Mass on the World" is primarily a "prayer," which Teilhard "prayed daily when unable to offer Mass" (*Teilhard's Mass*, 2005, xiii).

26 In *The Divine Milieu* this "death-force" is associated with what Teilhard terms "passivities of diminishment," both external ("all our bits of ill fortune") and internal ("the darkest element and the most despairingly useless years of our lives") (Teilhard de Chardin, *Divine Milieu*, 1960, 81). For him even "our failures, our death, our faults" can "be recast into something better and transformed" in God (136).

27 Boff develops this theme of cosmic divinization in *Cry of the Earth, Cry of the Poor*: "We must transcend the anthropocentrism that is common in Christologies, for Christ has divinized and liberated not only human beings but all beings in the universe" (Boff, *Cry of the Earth*, 1997, 174).

28 In 2017 the Pontifical Council for Culture unanimously requested that Pope Francis rescind this *monitum*, especially in light of recent popes' "explicit references" to his works in various speeches and documents, including *Laudato Sí* (see O'Connell, "Will Pope Francis Remove . . . ?," 2017). As of September 2019, the *monitum* remains in effect.

29 Emphasis in the original.

30 Lake is most perceptive on this point (Lake, *Incarnational Art*, 2005, 234): "Teilhard actually argues that Christianity is the only religion that promotes unity in which no particularities are lost. Pantheism is his example of a 'seduction' that emphasizes unity at the expense of particularity." In contrast to such pantheism, Teilhard's position here might

usefully be labelled panentheistic. Panentheism understands the divine as interpenetrating every part of the universe or "divine milieu" while simultaneously transcending it. Boff perceives that Teilhard's evolving cosmos "invites us to undergo the experience that underlies panentheism: at the slightest manifestation of being, at each moment, in every expression of life, intelligence and love, we are facing the Mystery of the universe-in-process" (*Cry of the Earth*, 1997, 154).

31 On the shared "theological aesthetic" of "pied beauty" in O'Connor and Hopkins, see Bosco, "O'Connor's 'Pied Beauty'," 2017, esp. 103–104.

32 In *Laudato Sí* Francis quotes this statement from John Paul II's 2003 encyclical, which stresses the "cosmic character" of the Eucharist. John Paul II does not refer explicitly to Teilhard, but this notion is unmistakably Teilhardian (Francis, *Laudato Sí*, 2015, 168 [no. 236]).

33 This diabolical work might be understood in terms of the central image in Leo Marx's 1964 book *The Machine in the Garden*. On this, see Piggford, "Excavator," 2018, 94.

34 O'Connor's story stresses the damage being done to the natural world through technology and the physical and spiritual consequences of that damage. At the same time, the Christ-like qualities of the natural world also evoke suffering humanity's need for relief. This comports well with Leonardo Boff's approach to integral ecology from the perspective of liberation theology. For Boff, to hear the "cry of the earth" is to hear the "cry of the poor," and vice versa: "To hear these two interconnected cries and to see the same root cause that produces them is to carry out integral liberation" (Boff, *Cry of the Earth*, 1997, 112). Pope Francis refers to these dual cries in *Laudato Sí* (Francis, 2015, 40 [no. 49]).

35 Pope Francis associates such a "habit of attention" with the "gaze of Jesus": "The Lord was able to invite others to be attentive to the beauty that there is in the world because he himself was in constant touch with nature, lending it an attention full of fondness and wonder" (*Laudato Sí*, 2015, 75 [no. 97]).

36 This scene anticipates a comment by Thomas Berry about an early perception of a meadow in his native North Carolina: "The more a person thinks of the infinite number of interrelated activities that take place here, the more mysterious it all becomes" (Berry, *Great Work*, 1999, 13).

37 Mary is also described as "mechanically" battering her grandfather, a term evocative of the work of the excavating machine.

38 Pope Francis reminds his readers of the "little way" of St. Thérèse of Lisieux as a hopeful mode of being that might provide an alternative to such a self-destructive course: "An integral ecology is . . . made up of simple daily gestures which break with the logic of violence, exploitation and selfishness. In the end, a world of exacerbated consumption is at the same time a world which mistreats life in all its forms" (*Laudato Sí*, 2015, 163 (no. 230).

39 This coda is included in an editor's note. On Fortune's secret will, see O'Connor, *Complete Stories*, 1971, 337.

40 Pitts himself is no admirable figure, especially in light of his abusive relationship towards Mary (see O'Connor, *Complete Stories*, 1971, 344). Mary's surviving parents and siblings, however, do at least appreciate the undeveloped woods. Although Mr. Fortune has sold the "lawn" to the devil-like Mr. Tilman for a gas station, the woods themselves will likely be passed along through Fortune's lawyer to the Pitts family (352, 337).

References

Benedict XVI. *Address to the Diplomatic Corps Accredited to the Holy See for the Traditional Exchange of New Year Greetings*. Vatican City: Libreria Editrice Vaticana, 2007. http://w2.vatican.va/content/benedict-xvi/en/speeches/2007/january/documents/hf_ben-xvi_spe_20070108_diplomatic-corps.html.

———. *Homily of His Holiness Benedict XVI, Cathedral of Aosta*. Vatican City: Libreria Editrice Vaticana. 2009. http://w2.vatican.va/content/benedict-xvi/en/homilies/2009/documents/hf_ben-xvi_hom_20090724_vespri-aosta.html.

Berry, Thomas. *The Great Work: Our Way into the Future*. New York: Three Rivers Press, 1999.

Boff, Leonardo. *Cry of the Earth, Cry of the Poor*. Translated by Phillip Berryman. Maryknoll, NY: Orbis Books, 1997.

Bosco, Mark. "O'Connor's 'Pied Beauty': Gerard Manley Hopkins and the Aesthetics of Difference." In *Revelation and Convergence: Flannery O'Connor and the Catholic Intellectual Tradition*, edited by Mark Bosco and Brent Little, 99–117. Washington, DC: Catholic University of America Press, 2017.

Buell, Lawrence. *The Environmental Imagination: Thoreau, Nature Writing, and the Formation of American Culture*. Cambridge, MA: Harvard University Press, 1996.

Davis, Doug. "Grace in the Machine: Technology and Transfiguration in Flannery O'Connor's Short Fiction." *Flannery O'Connor Review* 7 (2009): 18–34.

Delio, Ilia. *Making All Things New: Catholicity, Cosmology, Consciousness*. Maryknoll, NY: Orbis Books, 2015.

de Lubac, Henri. *The Faith of Teilhard de Chardin*. Translated by René Hague. London: Burns & Oates, 1965.

Duffy, Kathleen. *Teilhard's Mysticism: Seeing the Inner Face of Evolution*. Maryknoll, NY: Orbis Books, 2014.

Flanagan, Christine. "From Earth to Eternity: Ecocritical Approaches to 'Greenleaf' and 'A View of the Woods'." In *Short Fiction of Flannery O'Connor*, edited by Robert C. Evans, 159–173. Amenia, NY: Salem Press, 2016.

Francis. *Laudato Sí: On the Care for Our Common Home*. Frederick, MD: The Word Among Us Press, 2015.

Francis of Assisi. *Francis of Assisi—The Saint: Early Documents*. Vol. 1. Edited by Regis J. Armstrong. New York: New City Press, 1999.

Gooch, Brad. *Flannery: A Life of Flannery O'Connor*. New York: Little, Brown, 2009.

Graybill, Mark S. "O'Connor's Deep Ecological Vision." *Flannery O'Connor Review* 9 (2011): 1–18.

Grim, John. "President's Corner." *The Teilhard Perspective* 49, no. 1 (2016): 1–2.

Haddock, Keith. *Giant Earthmovers: An Illustrated History*. Osceola, WI: MBI Publishing, 1998.

Hopkins, Gerard Manley. *The Major Works*. Edited by Catherine Phillips. New York: Oxford University Press, 2002.

John Paul II. *Letter of His Holiness John Paul II to Reverend George V. Coyne, S.J., Director of the Vatican Observatory*. Vatican City: Libreria Editrice Vaticana, 1988. http://w2.vatican.va/content/john-paul-ii/en/letters/1988/documents/hf_jp-ii_let_19880601_padre-coyne.html.

———. *Message for the Celebration of the World Day of Peace*. Vatican City: Libreria Editrice Vaticana, 1990. http://w2.vatican.va/content/john-paul-ii/en/messages/peace/documents/hf_jp-ii_mes_19891208_xxiii-world-day-for-peace.html.

King, Thomas M. *Teilhard's Mass: Approaches to "The Mass on the World"*. Mahwah, NJ: Paulist Press, 2005.

Kinney, Arthur. *Flannery O'Connor's Library: Resources of Being*. Athens, GA: University of Georgia Press, 1985.

Lake, Christina Bieber. *The Incarnational Art of Flannery O'Connor*. Macon, GA: Mercer University Press, 2005.

Murphy, Michael P. "Breaking Bodies: O'Connor and the Aesthetics of Consecration." In *Revelation and Convergence: Flannery O'Connor and the Catholic Intellectual Tradition*, edited by Mark Bosco and Brent Little, 51–77. Washington, DC: Catholic University of America Press, 2017.

O'Connell, Gerard. "Will Pope Francis Remove the Vatican's 'Warning' from Teilhard de Chardin's Writings?" *America*, 21 November 2017. www.americamagazine.org/faith/2017/11/21/will-pope-francis-remove-vaticans-warning-teilhard-de-chardins-writings.

O'Connor, Flannery. *The Complete Stories*. New York: Farrar, Straus, and Giroux, 1971.

———. *Letters of Flannery O'Connor: The Habit of Being*. Edited by Sally Fitzgerald. New York: Farrar, Straus and Giroux, 1979.

———. *Mystery and Manners: Occasional Prose*. Edited by Sally and Robert Fitzgerald. New York: Farrar, Straus and Giroux, 1969.

———. *The Presence of Grace and Other Book Reviews*. Edited by Carter W. Martin. Athens, GA: University of Georgia Press, 1983.

———. *The Violent Bear It Away*. New York: Farrar, Straus and Giroux, 1960.

———. *Wise Blood*. New York: Harcourt, Brace & Co., 1952.

Paul VI. *Mysterium Fidei*. Vatican City: Libreria Editrice Vaticana, 1965. http://w2.vatican.va/content/paul-vi/en/encyclicals/documents/hf_p-vi_enc_03091965_mysterium.html.

Pertile, Lino. "Introduction to *Inferno*." In *The Cambridge Companion to Dante*. 2nd ed., edited by Rachel Jacoff, 67–90. Cambridge: Cambridge University Press, 2007.

Piggford, George. "Flannery O'Connor's Excavator and Dante's Lucifer in 'A View of the Woods'." *Flannery O'Connor Review* 16 (2018): 93–110.

Sadowski, Ryszard. 2016. "Inspirations of Pope Francis' Concept of Integral Ecology." *Seminaire* 37, no. 4 (2016): 69–82.

Scheid, Daniel P. *The Cosmic Common Good: Religious Grounds for Ecological Ethics*. New York: Oxford University Press, 2016.

Slovic, Scott. *Seeking Awareness in American Nature Writing*. Salt Lake City, UT: University of Utah Press. 1992.

Srigley, Susan. "Asceticism and Abundance: The Communion of Saints in *The Violent Bear It Away*." In *Dark Faith: New Essays on Flannery O'Connor's the Violent Bear It Away*, edited by Susan Srigley, 185–212. Notre Dame, IN: Notre Dame University Press, 2012.

St. John, Donald P. *Contemplation and Cosmos: Merton on Maximus and Teilhard. Teilhard Studies* 62. Woodbridge, CT: American Teilhard Association, 2011.

Teilhard de Chardin, Pierre. *The Divine Milieu*. Translated by Bernard Wall. New York: Harper & Brothers, 1960.

———. *Hymn of the Universe*. Translated by Bernard Wall. New York: Harper & Brothers, 1960.

———. *The Phenomenon of Man*. Translated by Bernard Wall. Rev. ed. New York: Harper & Row, 1968.

———. *Writings in Time of War*. Translated by René Hague. New York: Harper & Row, 1968.

Third Plenary Council of Baltimore. *Baltimore Catechism (A Catechism of Christian Doctrine)*. Number 3. Charlotte: TAN Books, 2010.

Vande Brake, Timothy R. "Thinking Like a Tree: The Land Ethic in O'Connor's 'A View of the Woods." *Flannery O'Connor Review* 9 (2011): 19–35.

Watkins, Steven R. *Flannery O'Connor and Teilhard de Chardin: A Journey Towards Hope and Understanding About Life*. New York: Peter Lang, 2009.

Westling, Louise. *Sacred Groves and Ravaged Gardens: The Fiction of Eudora Welty, Carson McCullers, and Flannery O'Connor*. Athens, GA: University of Georgia Press, 1985.

Woodward, C. Vann. *The Burden of Southern History*. 3rd ed. Baton Rouge, LA: Louisiana State University Press, 2008.

12 "Can You Make This All Run Again?" The Art and Environmentalism of Margo and Rein Vanderhill

Samuel Martin

Across the street from the college where I teach sits a small Episcopal house church with a bright red door. In one of my earliest memories of this wee church, I am leading a writing workshop in its sanctuary, once a living room. The workshop is focused on "writing with images," and we are surrounded by hand-painted Byzantine icons, hoping for inspiration. A haggard John the Baptist looks down on our manuscripts, as does a bloodless archangel and a dour Virgin. A life-sized Christ sits enthroned, severe of face and flanked by the four Gospel writers, all emblazoned by a sun-bright background.

A fifth figure stands perched on a ladder. He's a tall man with an unruly halo of white hair, dressed in shabby, paint-spattered clothes, and he's colouring in Christ's cheeks, like a New York make-up artist. Occasionally he chimes in on our discussion of *ekphrastic* poetry:

"You don't mind my two cents, do you?"

And I tell him his input is welcome; after all, the class chose to meet in this place hoping for some sort of divine inspiration. I nod at his ladder and he laughs.

"From them, I hope," and he motions to the icons. "Not me."

"We'll take what we can get where we can get it," I say, and introduce him to the students as Rein Vanderhill, a former art professor at our college.

Rein is likely to laugh at the idea of him being a conduit for divine inspiration, but he and his spouse Margo—both of them fine artists and now friends of mine—*have* in my seven years in Iowa borne witness to God's love for creation. In their work, in their lives, in the way they care for our small church, they have shown me environmentally conscious, creative ways to love a place and its people, even when that place seemed to turn on them in the fall of 2018 in a flood that nearly destroyed their house and over 50 years of collected artwork.

Given that the first paintings of Rein's that I encountered were icons, I assumed at first blush that he was a painter of religious subjects. But I was wrong. Raised in the Reformed tradition, Rein's chosen subject matter is the natural world: in particular the flora and fauna of the Iowa fields and Michigan shorelines where he and Margo have lived. Margo's artwork brings to life similar subject matter, but their artistic styles and intents are distinct.

DOI: 10.4324/9781003366744-12

Figure 12.1 "End of the Field," by Margo Vanderhill. Oil pastel on paper, 2016. Courtesy of the artist.

For instance, Margo's artwork in our little church is not painting icons but (among other things) furnishing the buffet that serves as the Eucharist's tabernacle with elaborate bouquets of wildflowers picked from the prairie behind their house that stretches to the nearby banks of the Floyd River. Margo's flowers, along with Rein's icons and the Advent candles' December light, help make this church a home for me and my family.

This sanctuary is one of the few places where I actually feel at home in this foreign country where I am—by birth and passport—a resident alien. I've found Margo and Rein's artwork helpful in my struggle to think of Sioux County as my home. By "home" I mean both grasping and giving away the otherness (or wilderness) of a place, which is how Don McKay describes "home" in *Vis à Vis: Notes on Poetry and Wilderness* (Kentville: Gaspereau Press, 2001). Home, for McKay, is not so much a building as it is holding one's fraught sense of belonging tentatively, loosely. Such a home has its beauty but also its terror. It can be lonely there as well as sublime.

Admittedly, I do not always think of Iowa, with its factory farms and endless corn and soybean fields, as sublime. When I think of encountering the sublime, I think of that sick thrill when I stood on a Newfoundland cliff-edge—a storm blowing in off the sea—and stared down at the rock-frothed waves, wanting to jump but knowing to do so would be death.

In Sioux County, the sublime is harder to perceive. Here, as a white Protestant male, I feel a strong pull to plunge unthinkingly into the quaint beauty of this place's small town Americana: to forget that though the place is named Sioux County, the seven Sioux tribes—the Mdewakanton, Sisseton, Teton, Wahpekute, Wahpeton, Yankton, and Yanktonai, among numerous other tribes who have called this region home—have been forcibly relocated to reservations south and west of here in order to create an "unsettled" Midwest to be sold to European and American settlers. Those early settlers, in turn, introduced to the region farming practices that have, over time, evolved into the modern factory farm, responsible in many ways for soil erosion, the pollution of natural waterways, and a host of other environmental issues fast becoming disasters. For me, not to heed such realities in talking about this place and its relation to my friends' artwork is to choose a Romantic sense of belonging over a realistic one: a false sense of unity with this place that further erodes and exploits its wilderness.

By "wilderness" I mean what McKay describes as "the capacity of all things to elude the mind's appropriations" (*Vis à Vis*, 2001, 21), the enduring strangeness—the otherness—of a particular place. McKay writes that the Romantic impulse "relieves us of our loneliness" (27). Romanticism tempts me because I do not want to be lonely here. I want to feel like I fit, like I belong. But the truth is I am lonely here, and should be. I do not belong here. That is the truth: that is real, and reality—realism as opposed to Romanticism—allows me to embrace my loneliness *in* this place as a reminder of my responsibility *to* this land and its colonially displaced Indigenous stewards. In contrast to the Romantic (or settler) impulse, McKay offers the option of "poetic attention," which recognizes, respects, and values "the other's wilderness" (*Vis à Vis*, 2001, 26, 28). Since the focus of this collection of essays is the environment, I'll limit my use of "the other" to mean a particular *place* (McKay uses the term more broadly), and instead of "poetic attention" I'll refer to the ways in which the Vanderhills practice "artistic attention." Such attention, according to McKay, can make possible "a *translation*" as opposed to "a *vestige*" of this place (28)—a flawed, human attempt to see more truly a part of the whole, since the whole cannot be grasped. In Margo's work and in Rein's, you can see this kind of artistic attention *to* and translation *of* this place—their home of many years. I see it in Margo's nocturnal townscapes and big sky vistas, in their blown-up details of the regions' botanical beauty in watercolour and acrylic, oil paint and oil pastel.

In order to make a home *in* and *of* this place, Margo and Rein have sought ways to acknowledge, bear witness to, or otherwise translate what makes them feel at home—the prairie flora and fauna, the magnolias and service berry blossoms Rein magnifies on canvas—as well as that which unsettles them. Rein's son Dietrik once pointed out to a mutual friend the fragility revealed in the ragged edges of his father's flowers. And Rein has told me that he paints botanical beauty as a coping activity to counter his anxiety and depression—"Anxiety," he says, "that is exacerbated by the destruction of nature I see in Sioux County, the most cultivated county in the state of Iowa," which is the most cultivated state in the United States. I think Dietrik is right that the pain of such anxiety can be seen in the ragged edges of Rein's flowers, in their obvious fragility, and in the darkening depth of many of his compositions.

Magnification, ragged edges: such elements, to my eye, defamiliarize the familiar so that, in one of Rein's tryptichs, wild magnolias, blown across three canvases, become a "Flight of Angels." McKay writes that in such "defamiliarizations, often arranged by art, we encounter the momentary circumvention of the mind's categories to glimpse some thing's autonomy—its rawness, its *duende*, its alien being" (*Vis à Vis*, 2001, 21). The literal translation of *duende* from Spanish is "ghost," "demon," or "goblin"—suppressed or subterranean things that enchant and haunt—and there is something that enchants, that haunts Rein's work, and Margo's too: something uncanny in the chiaroscuro of light and shadow, the contrast of homely and unhomely, warm welcome and cold shoulder. A marked ordinariness makes Rein's angelic magnolias fragile and makes me wonder not whether these flowers are as beautiful as angels but whether angels—whatever they might be—are as beautiful as these magnolias. And when I turn to Margo's small-town nightscapes, I spy a home's interior light that simultaneously alienates and welcome me—an artistic rendering of what I have found to be the confusing complexities of Midwestern hospitality.

If the hospitality of Midwesterners confuses me, the hospitableness of the Midwest's environment does so to a greater extent. Gardening, for instance, is almost effortless here, especially for my wife who remembers rock-picking her parents' rough farm fields outside Ottawa, Ontario, the air bloody with blackflies. But though gardening is easy in Sioux County's "black gold" soil, I have never lived in a place where I am as constantly aware—and afraid—of what extreme weather can do, whether by tornado, hailstorm, heat, or flood. As a result, I feel vulnerable to the weather in a way that might make an Iowan smirk.

I remember hearing my first tornado siren while standing in line at a local coffee shop. I stepped out of line immediately and buzzed straight for the building's basement, hearing my buddy Gary laugh. "This your first twister?" he called. And I thought then that nothing could faze a weather-hardened local like him. Margo and Rein moved to Sioux County in 1974 and have weathered forty-six years in the region, the last eighteen spent in a house in Alton that they built on the edge of the Floyd River's flood plain. In that time they've only ever experienced two or three floodings of the low, wet mesic part of their backyard prairie, but, as Margo notes, "nothing threatening the house because the dyke in the backyard—between house and prairie—has been enough of a protective wall."

The spring of 2018, however, was wetter than usual and the Floyd flooded in June, leaving only eight inches to spare at the top of the Vanderhills' dyke. The rains continued through the summer, and the soil became so saturated that when up to six inches of rain fell on Wednesday and Thursday of the third week of September, there was no place for the water to go. Unlike earlier that summer, no sheriff's deputy stopped by to alert the Vanderhills of flooding. And according to all the signs they'd learned to read—both out their window and online—this flood looked like it would be "a quieter one."

The Vanderhills still kept watch, though, checking the extra sandbags on their dyke and the NOAA weather maps and constantly refreshing the Advanced Hydrological Prediction Service App online, which indicates the Floyd's local water level. During the summer flood, months earlier, Margo and Rein had hauled heavy items up from their basement, but given the readings for this flood and the extra height added to their dyke, they thought simply getting things off the basement floor would be enough. "I thought some water would come in," Margo says, "so I put boards on the top of the tub, and baskets of laundry on the boards. We even had our two large, newly purchased squeegees to direct any water to the floor drains, thinking our sump pump could handle it."

But the water soon began streaming over the road, making the drive to the bridge a fishing ground for the local great blue heron. And then another band of rain appeared on the weather map, a storm predicted to arrive at the same time the river was supposed to crest.

Margo remembers looking out to see a road barricade swept across the flooded prairie toward the creek. They found the pieces on the creek's far side weeks later. At that moment, though, they knew they'd need more sandbags, so they raced in their truck past the golf course upstream from them, where kayakers were paddling the course. By the time Margo and Rein returned home with their 20 sandbags, though, water was coming through the bags already in place and the wind was whipping waves across what had been prairie the day before.

They placed the 20 bags and filled 30 more to try and complete a second row on top of the dyke, but the bags could only be moved three at a time in their wheelbarrow. It was slow, hard work but they got the 50 bags in place and Rein wondered if they'd beaten the rising water, now significantly higher than earlier predictions. The rain stopped and they thought that might be it. But then at 6:30 p.m. a fast-moving line of thunderstorms drove more rain across the flood zone. As the storms whipped through, Margo and Rein could see the water visibly rise and begin flowing over their uneven line of sandbags into their backyard.

They quickly sandbagged their basement's double doors, the water already ankle deep. Then they rushed to the basement and began lifting more items off the floor onto workbenches, high shelves, and the table saw. They stowed things as quickly as possible, without priority or plan: both of them panicked, fearing the water would rise beyond the sandbags walling in the basement door. When the water did rise that high, Margo grabbed a floor squeegee to direct the water through the sunroom to the drain in the workshop floor, but with a gunshot bang the doors burst open and the river began gushing into their basement. Margo remembers grabbing the door and trying to push against the arcing flow while Rein rushed around the workshop yelling "O my God! O my God!"

In the wet chaos, they found each other in their workshop and saw their refrigerator and freezer come to life like awkward, dancing robots, shifting side to side before upending to bob door-side up. They left their appliances floating like that for several days, sealed like Queequeg's coffin at the close of *Moby Dick*, their food preserved inside, untouched by the polluted river water, which was black with silt, and whatever else the nearby town was venting into the river to ease pressure on their insufficient storm drains.

While Rein stubbed his toes wading room-to-room in the waist-deep water, Margo splashed past their tax records on a counter in their studio, grabbed a few files of contact information and her computer, entangled with router, speakers, and other wires, and hauled the lot to the top of the stairs, only to bump the pile a few moments later and send the computer bouncing back into the black water below.

Margo tells me that later they did manage to save the computer's hard drive—"Thank God"—but in that moment all they could think to do was to haul the computer back up the stairs and continue salvaging a few more items from the studio, noting the seaweed-like tickle of submerged magazine pages on their sandaled feet and the strange calm as the water stilled.

"Small items gently bobbed in the water," Rein says. "We found a dead mouse floating, then two more live ones, each on its own flotsam. We scooped them up with a cut-off milk jug and released them onto dry land. Fish schooled near the mostly closed door, sensing light or current. But what did *we* feel? Resignation and shutdown. We retreated to the upper level of the house." It was now 7 p.m., the light fading. Rein noticed a fluorescent glow in the basement, so he waded back into the silty water, pushed the switch, and got a shock, noting at the same time a certain tingling in his legs. *The water is charged,* he thought, realizing most of the basement's outlets were submerged and the breakers had not yet tripped. With rubber gloves tossed to him by Margo, he pulled the main breaker.

Silence and darkness.

Rein's electric experience ended their work for the night, and they went upstairs to eat by candlelight. "It was a night of incongruity," Margo says. "The upstairs was in order, except for the lack of electricity. We have a gas stove so we could cook. But the basement was like Venice, lapping water inside the house, dark cellars full of water. I remember lying awake to a mysterious gurgling below."

"The wall cavities filling with water," Rein says.

The next morning, they found that the water had dropped about 18 inches. By 4 p.m. the level was about two feet and dropping very slowly until Dietrik came with a two-inch motor-driven trash pump and sucked out the remaining water in about an hour, leaving a muddy sludge on the floor. The furnace was a loss, but Dietrik revived the dryer and washing machine, though the latter's rinse cycle now sounds like a jet engine.

Once the water had been pumped out of the basement Margo and Rein began hauling wet, dripping loads out to their driveway, spreading woollen coats and other clothing out on the concrete in rows, sprawling acrylic paintings on wooden shelves, clustering art materials and tools and small sculptures under the tree on the front lawn. It didn't matter that they were exhausted, they forced themselves to keep working in order to stay ahead of the mold that was sure to bloom quickly in their wet basement.

~

McKay writes that we "probably experience" wilderness "most often in the negative as dry rot in the basement, a splintered handle, or shit on the carpet" (*Vis à Vis*, 2001, 21). Add to that list all the ruined furniture and the warped and swollen doors and drywall that Margo and Rein lugged out of their basement and tossed in a 22-foot roll-off dumpster. Or every single screw, bolt, or electrical part in the more than one hundred drawers in their workshop, all of which had to be sorted and either discarded or scrubbed clean of river silt.

When I visited the Vanderhills' house months later, Rein showed me a back basement room in which, before the flood, he had created a miniature Midwest, an elaborate toy world in which model trains wend through recognizable hand-crafted hills and towns, all built at chest height—the same height the water reached before receding. As I stood with Rein in that room, odd Gullivers in an all-American Lilliput, I thought of Margo's recollection of lying in bed in their untouched upstairs, listening to the basement's gurgling as water seeped through drywall and filled the spaces between wall joists.

Months later, Rein would show the trains to my two boys, and Liam, the eldest, would ask, "Hey, Rein! Can you make this all run again?" An honest question because though the painted topside of Rein's toy world remained dry, the elaborate circuitry on the underside of his chest-high tables still needed to be cleaned of river silt before the trains could run again around the room. In that corroded circuitry, I'm sure Rein felt, in miniature, the larger encroachment of wilderness on his life—a tingling like the electrified water in which he'd once been shocked.

Shock is a complicated experience.

For the Vanderhills, there was the *initial shock* that sent Rein running around the room, knee-deep in water, yelling "Oh my God!" or the shock that drove Margo to try to heave shut a door slammed open by the force of a raging river. Then there is the numbness of being *in shock*, just doing what had to be done, no matter that they were in their seventies, dog-tired, traumatized, and emotionally spent. Add to that the *aftershock* of starting awake in the middle of the night two days after the flood to the clanking crash of a full-sized train as it derailed off a nearby bridge and threw its 37 cars into the Floyd River, all because the flood had washed out the rail bed. Margo tells me that the crew escaped injury and, thankfully, no toxic chemicals were spilled into the river.

I try to imagine the screeching, industrial grind of the crashing train in the night. Raised in a Christian tradition that spoke often of the Rapture—the sudden snatching of Christ's faithful from the face of the faithless earth—I likely would've shot awake thinking it was the end of the world—a dark Hollywood apocalypse. We often think of apocalypse as the end of the world—some stark cataclysm—but an apocalypse, in the Christian tradition, is not necessarily the scorched earth of Cormac McCarthy's *The Road*; rather, it is a revealing of that which has been submerged or forgotten, settled. In the Christian tradition, apocalypse unsettles that which is settled, just as the coming Kingdom of God upsets the regimes of this world, including religious systems that under the auspices of law, purity, and holiness, mark borders the faithful are forbidden to cross, because beyond those borders lurk the unlawful, the defiled, and the unholy. The polluted, lepers, demoniacs, Samaritans, and tax collectors were, in Jesus's day, among the many deemed "unclean" by the religious leaders who kept to the Pharisaical purity code. Yet Jesus declared those unclean clean by forgiving them their sins, healing them, and including them by eating and drinking with them or calling them to be his disciples. Jesus made clean these unclean people, according to the Gospels, as signs of his coming Kingdom, revealing that at the heart of the Kingdom of God is God's own love for the *anawim* or as Terry Eagleton translates the term, "the scum of the earth." The unloved. In this sense, Jesus's ministry was apocalyptic.

But what did the flood reveal to Margo and Rein? What did this disaster allow them to see? In their clean-up, they emptied flat file drawers crammed with drawings, dumped silt-destroyed pieces, and displayed recoverable art in their garage, where three dehumidifiers and four box fans whirred constantly. The strange result of this "triage-salvage," according to Rein, was a disaster-zone retrospective spanning decades. "The experience," he says, "was a positive one." He cuts himself off with a sharp laugh. Runs a hand through his shock of white hair. Shakes his head. "Minus the destruction of much of our art, of course."

For Margo, on the other hand, sorting through their soggy artwork was intense. Many things were ruined, yet she was quite happy to get rid of the mediocre and never-quite-resolved work. "Everything was quite fragile," she admits. "Rein's watercolors could be cleaned, but at the loss of their intensity. My landscape sketches in ink transferred to other drawings, creating layered landscapes." And the Floyd's silt, when it didn't obliterate a piece, often added a subtle smoky wash or cloud, altering the piece's value and interest unpredictably.

According to Rein, everything about the flood felt unpredictable. "I'm the more anxious one," he says, "and now experience an almost PTSD reaction when heavy rain occurs. Even though we have added two feet to the dyke separating us from the flood plain and have purchased flood insurance, I still have an autonomic adrenalin surge, a fight-or-flight response to rain." Rein's aesthetic disposition has been nurtured in this place, in this rare oasis of prairie in a factory-farmed state. "We've had eighteen years of intimacy with the natural world," he tells me. "But now three floods in twelve months. I at least feel some abuse. As if a spouse has suddenly become violent. I tend to anthropomorphize the river more now, even though my science brain knows it moves with no mind of its own."

During the height of the flood, Rein remembers seeing a rabbit swimming to higher ground. "I had never seen a rabbit swim," he says, "and it made me realize that the flood had not just affected us or our human neighbours, but the whole ecosystem in which we live." After the flood, Margo and Rein went to dismantle a neighbour's lumber pile that had washed onto their property, but discovered a mink's lair inside, with two frogs laid out carefully like a fisher's strange catch. They left the lumber pile as it was, again realizing the flood's differing effects. Whereas some houses were destroyed, animal and human, some, like the lucky mink's, were newly formed out of the seething chaos.

Rein has told me more than once that if the flood has revealed anything, it is just how seamlessly humans are part of the muddy earth on which we build our homes: merged with the evolved continuum of that which walks, creeps, and crawls on the earth. "The flood," he says, "has deflated my delusional sense that I have of control over my life or environment. Now I live with a more realistic anthropological anxiety." He points to the painting by Titian called "The Sacred and Profane," in which two women pose on each side of a fountain, one clothed and one nude. "The nude," Rein says, "is the sacred, un-altered by humans, straight from God—wild, unindustrialized prairie. But I see the agriculturally-altered earth, particularly here in Sioux County, as profane. For me, the natural aesthetic of the almost infinite diversity of the nonhuman landscape is sacred. Untouched prairie. Which is why Margo and I have attempted to reconstruct, as best we can, a wild prairie between our house and the river."

Rein captures a small detail of the biodiversity found in this thin strip of land in his painting called "A Prairie Perspective." In the composition, Great Blue Lobelia grows next to False Boneset and Canadian Rye shares roots with Bottle Gentian, New England Aster, and onion; Big Blue Stem, Brown-eyed Susan, and Indian Grass arc into Smooth Aster; and Purple Coneflower heads darken the foreground. The scene comes straight out of their prairie, Rein tells me, in which they observe "the symbiotic nature of the prairie's diversity. Over 125 species of forbs and grasses coexist in a balance sustained by fauna in the earth and air: a profound symphony of difference and community. A model for humans to do the same, to pursue a mutually beneficial cohabitation."

For Margo, art and prayer can by times form a similar cohabitation. For her, visual art can take on the quality of a prayer. "Sometimes," she says, "my art is a form of praise, a hymn celebrating the beauty of creation. At other times, it's not so much prayer as a search for an emotional element, such as my domestic night scenes, which explore the intimacy of interior space, lit, cozy, but closed to the outsider—to me standing alone in the street. The question I keep coming back to is: Who will be given hospitality?" An apocalyptic question to which Margo's artwork takes her back repeatedly.

But her creative work that more directly addresses environmental degradation goes beyond what she does in visual art, she says, and includes their reconstructed prairie and the more than 80 trees they've planted that now provide habitat, soil erosion control, and the photosynthetic means to produce oxygen, all of which help preserve the kind of grassland ecosystem fast disappearing from this part of the world.

And what happens when these grasslands disappear?

Dr Laurie Furlong, a friend of Margo's and a biologist at the local college, says that prairie grasslands preserve and create "fluffy topsoil" and that modern agricultural practices contribute to the soil's degradation, to the dirt losing its "fluff." Deep-rooted perennial prairie grasses, like those Margo plants and draws, grow and decay and make the soil "fluffier" with their partially decomposed organic matter. "You can feel this when you walk through a tilled garden," Laurie tells me, before going on to say that we now have a different ecosystem in Iowa, one composed of short-rooted annuals, like corn, which contribute little to soil building. "Add to that," she says, "the liberal application of inorganic fertilizers and pesticides that interfere with decomposition, soil that is left bare to rain and snow, and tractors that further the damage through compaction." Simply put, the soil is not as "fluffy" as it once was—there are fewer spaces between soil particles and there is less organic matter in the soil. "As a result," Laurie says, "water from rain and snow-melt penetrate less deeply and runs off more readily. And as the runoff rushes over compacted soil into streams and rivers, these waterways overflow their banks, causing floods." To help slow water movement and soil erosion, farmers in this area have installed terraces in fields and buffer strips on stream banks. Unfortunately, Laurie points out, neither of these efforts contributes to soil *building* with the same effectiveness as the reconstruction of prairie grasslands, whose soil is enriched by the deep-rooted, native perennials of the kind tended by Margo and Rein in both their studio and backyard.

As Margo points out, though, these grasslands, which are so important to them, are fast disappearing in Iowa. Not all such grasslands have disappeared, of course, but they foreseeably could. And what then? What are we witnessing when we see the last acres of prairie plowed into farmland or industrial agricultural practices rendering soil barren or human-propelled climate change turning more and more of the Great Plains into floodplains? McKay writes that in like instances what we are seeing is not merely the land being "used" as a tool—an agricultural tool in this instance. What we are witnessing is the land being "used up," made into *matériel*: the endgame of colonization when the land can be neither farmed nor returned to prairie; the damning result of demanding higher and higher crop yields with less and less nutritional value until the soil can either give no more or temperatures rise too high for crops to fertilize.

That is, if the floods don't spread first.

For now, though, small plots of prairie still exist, in no small part due to the efforts of people like Margo and Rein. Their prairie, to them, is a joint work, both creative and educational. They are both, after all, teachers as well as artists, and they regularly open their prairie to elementary classes: to third-graders who are learning about seeds and how they are distributed. Those students collect seeds from the Vanderhills' grassland and plant them elsewhere in the area. Other groups occasionally visit: Brownies, high school and college biology classes, and adults. Like me.

"Occasionally," Margo says, "we give prairie seed to colleges and friends who are reconstructing prairies to heal environmental damage. Beyond our work on the land and educating about the importance of the environment, we installed the first solar panels in this area on our garage roof. We thought since our land was so close to a wildlife 'hotspot' that solar panels would be less harmful to the many birds in the area than wind turbines." These panels and the planting of their prairie represent small attempts to repair damage. However—and Margo is quick to point this out—personal action alone is *not* enough to prevent environmental degradation or climate change. Advocacy and collective action are also needed: community-wide creative endeavours of the kind the Vanderhills have pioneered in this area.

But such collective action is hard to spur in Sioux County, let alone sustain. The Vanderhills know the scepticism with which creation care can be met here; they've dealt with it for more than two decades. "Environmental practices in this area varies a lot farmer-to-farmer," Margo says. "Some build erosion-controlling terraces while others run their machinery too close to rivers. Judging from the soil in some places, some farmers must drag harrows and planters with one air-borne wheel spinning freely over the river." The resulting loss of soil is depressing and worrisome, Margo says. "The loss of an inheritance."

In her time in northwest Iowa, Margo has witnessed many of the small remnants of prairie erased by ditch dredging and dairy enlargements, too-frequent haying and spraying of herbicide, new farm management and highway construction. When she passes these places now, she grieves their loss. "Twenty years ago they were just small traces of the once Great Plains," she says, "and now they are all but gone. Sometimes I just want to look away, like I did when the county crudely lopped off large limbs from the trees around the courthouse. Maybe I shouldn't have. Maybe I am like Laurie Furlong, who once confided that she'd like to drop teaching environmental science because it made her feel depressed. Maybe we need to bear more pain. I don't know. I do know that living in Sioux County is not easy for those who see."

And even though I've only lived here six years, I see what she means. Of course, it's not only here; it's everywhere. It's global. And it's terrifying. I read books like Bill McKibben's *Falter* (2019), and I learn that if we continue to raise carbon dioxide levels, we may lose the ability to think straight. In an op-ed for *Rolling Stone*, McKibben writes that by the year 2100 human cognitive ability could fall by 21%, impairing, among other things, our capacity for crisis response, using information, and strategy-making—skills high on the list of necessities for human survival. I read stuff like that, and I do the calculations and realize 2100 is foreseeably within my children's lifetime, if they wind up having the same life-expectancy I enjoy.

Liam, now 6, will be 87 in the year 2100.

Micah will be 85.

And Charlotte will be 82.

And what of their children? Or their grandchildren?

Being a writer, one given to dark imaginings, I can visualize what's coming: the ways in which my children will experience first-hand the very things that the Bill McKibbens of the world predict for this small earth. Like Margo, sometimes I simply want to look away. And like Rein, the sheer scale of environmental degradation—even just in this county—leaves me depressed. Nearly debilitated.

How then can I learn, as I must, to bear more pain?

For McKibben the answer is art.

Art, he says, allows us to feel more fully the losses we suffer due to climate change, losses that conscientious environmentalism might try to mitigate or manage. But beyond deepening our capacity for empathy—or encouraging us to install windmills in our backyards, against which to tilt like environmentally-woke Quixotes—can art really do anything? After all, neither Rein's giant magnolias nor Margo's nighttime streets will stop another flood. But those angelic petals and burning-bush windows do seem to keep the Vanderhills rooted. Like the native grasses of their prairie, which they translate to canvas with careful artistic attention, using pencil and ink, oil and oil pastel, watercolour, and acrylic.

And though their art may not save the place they call home, their art does help *make* this place a home. To echo McKay (23), their art both grasps this place and gives it away. Each work clutches a fistful of flora, given in turn as a bouquet: an artful arrangement that alleviates some (though not all, never all) of the pain I feel as an alien in this country, an illegitimate guest on these Indigenous lands. That is something. Something tentative and fraught, which I try to hold loosely and with thanks. The Vanderhills' art also creates hope as surely as their prairie and woods create oxygen—hope bright as a lamp-lit window on a dark street, fragile as a flower tattered by reality's hard facts: my children's futures.

~

It is Advent now and I've just received a letter from the president of my college informing me that my position is being terminated. When I began this essay, I did not realize it might be my last as an English professor. Ice rain lashes the window of my office and I realize I have to walk home in the wet dark, past the little church with its bright red door—painted saints on the walls within. Gone is midsummer when Rein and Margo first walked me over their newly raised dyke, and gone is the autumn day when my wife and I picked apples with Margo and our baby Charlotte while Rein took our boys to see his model trains.

Now is the time when Christians recall Christ's first coming as a Middle-eastern refugee. Now is the season in which liturgical art and prophetic poetry help the faithful imagine a future in which Christ will come again in glory.

But what will that glory look like if our first glimpse was Jesus of Nazareth, if God's goblin desire from the very beginning—the divine's dark *duende*—was to be born homeless and hungry, hunted by a corrupt king? What kind of God is this? "Who is this king of glory?" we sing. Not a rebel messiah seizing power but a refugee-turned-rabbi bolted to a bloody cross. And what will his coming glory look like? Will we recognize this coming Christ in the faces of the climate refugees flooding our borders? Is that what the coming day of the Lord will look like?

And what will the earth undergo before that day arrives? What floods will my children face, and will they fare as well as the Vanderhills? To cast Margo's artistic question to our eschatological end: On that day, who will be given hospitality? And will we—finally, like the too few who now do—show hospitality to this *anawim* earth, our island home?

I don't know if it's learning that I've lost my job that has me in this mood, or whether it's a result of looking more closely at what's happening to the natural world around me, but as I write this, I stand in my mind with my sons and Rein. We're in Rein's room full of model trains, the tops of my boys' heads well below the flood's watermark. Had they been in that room when the river burst through Margo and Rein's basement door, they would have been drowned. The thought terrifies me, even now, as if *my* head is being held underwater. Yet a glimpse of Rein's red service berries or a wink from one of his stoic saints, a passing glance at one of Margo's pastel drawings or the peppery scent of a prairie bouquet—my kids crying, "These flowers are from Margo's prairie!"—and I am able to breathe. And I am filled with joy and hope—hope beyond all reason and of the kind in which Advent helps us believe.

I say this yet have no clue as to the power of that final apocalypse—Christ's ultimate *duende*. What will so captivate on the day of Christ's return that even the dead will rise? We say we will behold God's glory, but what does that mean? After the last glacier has melted and Lagos and Hong Kong and a thousand island cultures have gone the way of Atlantis, will the flood waters finally recede? Will

Figure 12.2 "Reconciliation," by Rein Vanderhill. Acrylic on canvas, 2019.
Courtesy of the artist.

we as a species climb again out of the mud and muck, silt and sludge more polluted than primordial?

Will we again be able to breathe and think clearly: to imagine better ways of being in the world? In that end—our new beginning—will our long, egregious, and utterly mundane list of environmental sins so easily be rolled back as a scroll?

Or will we on that day, like my boys in Rein's basement, stare up at God's sad face, at that tired smile, and at that shock of white hair we once took for a halo, and ask—as if anything were now possible, as if we'd done nothing wrong—"Can you make this all run again?"

References

McKay, Don. *Vis à Vis: Field Notes on Poetry & Wilderness*. Kentville: Gaspereau Press, 2001.

McKibben, Bill. *Falter: Has the Human Game Begun to Play Itself Out*? New York: Henry Holt, 2019.

———. "This Is How Human Extinction Could Play Out." *Rolling Stone*, 9 April 2019.

13 Environments of Grace

Reflections on Sacramental Reality in the Work of Bruce Cockburn and David Adams Richards

Norm Klassen

Introduction

The word "environment" is, in some ways, a near-synonym of "ontology." Both bear analogical relation to the Eucharist, the celebration at the heart of the Catholic Mass. We do not all agree about the meaning of what happens in the Lord's Supper (Christians with Christians, let alone Christians with those of other faiths or none at all), and this confusion penetrates our approach to the environment, since our interpretation of the Eucharist reveals our ontology, or view of the nature of reality.

Sacramentalism prepares the ground for effective environmentalism and provides the seeds of respect for the created order. It cultivates wariness about human projects, including environmentalism's own susceptibility to political posturing; however, the Mass, insofar as it is a human project, also affirms human cooperation with God and encourages humanity in the care of our common home. A sacramental approach to Creation, rooted in incarnational theology, was historically part of Church orthodoxy, especially in reaction to dualistic heresies that denigrated the creation and put God at a distance from it.

The features of a sacramental ontology illuminate environmental issues and encourage a certain response to them, but let me not overstate the case. Our (environmental) needs are many, and our world complex. A few theoretical considerations here are not going to be decisive; they do not even lead below to an action plan. My hope, though, is that this paper might help the reader to keep thinking about and imagining what a sacramental ontology entails and that they might find contemplating our environmental needs a fitting occasion for doing so.

This ordering may strike some as perverse—that the real object of concern around which we can *all* gather, surely, is the environment, not a religious belief. Christianity demands the subordination of even so lofty a concern—so ready-made an object for the setting aside of differences—to a right ordering of our obligations. The right worship of God takes precedence, and sacramental ontology means nothing if not a soul-searching reflection on disordered desires—the universal tendency to idolatry—within the human heart. Providentially, the right ordering of our desires normally leads one back to take up again noble and fundamental human ambitions, among which the proper care of our common home must rank at or near the top. Nonetheless, the ordering is absolute, and the expectation, providential

DOI: 10.4324/9781003366744-13

though it may be, must be suspended. Environmentalism begins in existentialism of a particular kind, and we must think away any intention to arrive by surreptitious means at what must always be, ultimately, a secondary goal. Not only do our souls depend on it, but, ironically, the environment as well.

Sacramental ontology reminds us of the givenness of all things. From this recognition follows with road-to-Damascus-like clarity the need to resist treating anything in the world in an instrumental way. Nothing in the world can be reduced to the uses to which we would put it. Environmentalism itself can deteriorate into self-propitiating and self-aggrandizing posturing, which pollutes political discourse and, ironically, utilizes the environment for its own human-centred purposes. (The tendency towards instrumentalization and legalism loom perpetually, but thank God for the provision made against them anteriorly from within the givenness of things and made manifest in that which the Mass re-enacts.)

In what follows, I will first sketch sacramental, participatory ontology. Then I will follow the lead of two Canadian cultural, literary theologians on how to resist an instrumentalizing attitude, with some early lyrics by Bruce Cockburn, who is more Catholic than he perhaps realizes, and the threat posed by environmentalist posturing, with some pages from an early novel by David Adams Richards, who is more orthodox as a Catholic than he perhaps intends.

Sacramental Ontology

Catholics believe that, in the Mass, the elements become the body and blood of the Lord.[1] Christ is really present. Through the mediation of the priest, the elements of ordinary bread and wine *participate* in the being of Christ in the most intense way of any created reality. At the same time, the elements indicate the true status of all reality. All reality participates in the reality of God.[2] All reality, moment by moment, derives its being from God. The Mass re-orders communicants to this truth in a weekly if not daily rhythm, and through them it extends out into the world. The faithful, gathering for the Mass, bear witness to the wider world in a twofold political act: in the testimony of a society attesting that which matters most to it and in the subsequent actions of the congregants now dispersed in the community like so much yeast.[3]

We sin, we forget—who but the saints can faithfully inhabit and live out the truth that all reality is sacred and that nothing can finally be treated in instrumental fashion?[4]—we confess, we return to the sacred mystery to be revivified by it. The priest plays an active role in this process in the consecration.[5] The people likewise enter into the mystery of Christ's sacrifice through their free assent, so that salvation is entirely the work of Christ on the cross and at the same time involves the people of God, who play the role to which Christ invites his disciples at the Last Supper, in the drama of redemption. Mary exemplifies that role, her *fiat*—even her willingness to let Christ suffer (as Peter at first could not)—symbolizing and embodying the obedience and life of the Church.[6]

There are certain other things that one could say that might seem to have more direct bearing on how one thinks about the environment in sacramental terms, that have been said, not least in Pope Francis's celebrated encyclical *Laudato Sí*. In one way or another, these insights all return to the Eucharist.

(1) There is no *natura pura*.[7] Christians wisely resist the use of the term "nature" in favour of "creation." Nature represents the depersonalization of the created order. It is meant to be a neutral term, but, like many terms used within the frame of secularity, it carries within it the seeds of an outlook that refuses to acknowledge givenness. Christian preference of the term "creation" amounts to the affirmation of God's existence, the reality of creaturely freedom, and the created order's dependency on Him. It is, however, still possible to restrict God's role to a deistic one: that he sets the world in motion, then leaves it to run on its own. Christian tradition has long resisted such dualism on the grounds of the incarnation, in combination with an ontology the Church shared with other premodern societies for the first thousand years of its existence. God has not left us on our own but is continually active and present in His creation.

In refusing deism, one can nonetheless suppose that God enters a world otherwise self-contained and knowable in scientific terms. The widely shared premodern ontology of participation includes dimensions not countenanced in a modern construal, which Christian doctrine both reflects and enriches. Words like *physis* and *natura* denote a synthesis of the divine, anthropic, and natural (in the modern sense) realms. (Dupré, *Passage*, 1993, 3–5). The incarnation simultaneously reveals Christ's coming and the revelatory nature of reality. It makes manifest all creation as caught up in the movement of Christ's going out from and returning to God.[8] All of reality was only ever to be made sense of in relation to self-giving and suffering. It yearns for its own fulfilment in the divine life that sustains it and models self-giving, suffering love.

There are at least two reasons why some Christians find sacramental logic challenging. First, the connection to Greek and Roman thought makes this reflection on the nature of the real seem philosophical at the expense of biblical revelation. Second, it appears to render less concrete the human reality with which Christ identifies in his incarnation: that is, turning away from nature as essentially something already known can seem to detract from the clear salvific message of the gospel. We thought we *knew* that to which he was coming to save. How else could we know the significance of his coming? Furthermore, why complicate a salvation story with so many ready-made points of application to the saving of the planet?

Yet a counter-question presents itself: does one secretly wish to preserve the autonomy of one's knowledge of the world in wanting to (over) simplify the description of that with which Christ came to identify? This latter question implicates not only the problem of *natura pura* but also the depravity of the human heart: our capacity for self-deception runs deep. There lurks in ontology and environmentalism alike the possibility of the subtle reinforcement of autonomy, whether the individualism of neoliberalism, the broad-spectrum autonomy of non-transcendence, or the creeping refusal to live by faith. The threat to what we know destabilizes our sense of orientation in the world, especially in a "knowledge-based economy"!

(2) Nature finds its completion in response to the divine. Humans play an exemplary role in this regard, at once unique and shared. They are not so much distinct from the rest of creation as they are leading it, through our elder brother Christ

(Heb. 2:11), back to God.[9] At the same time, they are uniquely gifted with reason and uniquely capable of self-transcendence. They are not enclosed by their natures; they can educate their desires. It begins in the Mass. They can submit themselves in obedience to God, they can sacrifice themselves for the other. This activity, paradoxically, makes it possible for them to be *more* themselves. They discover themselves in giving themselves away.

This activity is paradigmatic. What they do for themselves they do for the rest of the created order as well. This is why old-fashioned words like "husbandry" and "dominion" imply hierarchical and paternal care. To take care of plants and to nurture and protect animals is both to exercise God-given responsibility and to play an appropriate role in helping the other orders of creation to achieve the potential fullness of their own being in God.[10] In this time of the Anthropocene, we hear much about the ill effects of human presence in the natural environment, and we are rightly ashamed of our recklessness and our greed. We exploit and are disobedient where we should nurture and care.

A recent curated display at the Royal Ontario Museum in Toronto, entitled "Life in Crisis," illustrates this deleterious dimension of our activities, observing that "humans are causing the extinction of individual species and the destruction of entire ecosystems" (*Life in Crisis:* Schad Gallery of Biodiversity, Royal Ontario Museum, July 2018). Yet the human role in the natural order is not altogether negative. God did not make the world that way. Creation can only be what God intended it to be through the active role of people caring for the environment and with people—starting preferably with his chosen representative society on earth, the Church—modelling appropriate behaviour in advocating and taking up in continual fashion the biblical mandate. The Church has a massive and irreducible role to play in the preservation of the created order.

(3) Our involvement with all of creation is rational. We are often tentative in our responses to the world around us. We intervene most quickly in situations that elicit emotional reaction, but in emotivism there is little evidence of coherence in our interventions born of confidence in a shared vision deriving from practical wisdom or an affirmed tradition.[11] In certain quarters, belief in rationality has collapsed under the hermeneutics of suspicion. The necessary critique of Enlightenment rationalism has produced a one-sided and adventitious constructivism.[12] On the other side from either emotivism or constructivism, the bottom line, if not naked greed, dictates corporate and even governmental policies. Many people express disgust at market forces, but in their lived philosophy acquiesce to a creed of self-interest that not only blurs our responsibilities but imposes a canopy on the human potential to do good for others more suffocating than a report out of Davos.[13] We have lost belief in our greater capacity for reason, and not only because of the ravages of postmodernism.

Yet our real situation, our historical situatedness, is that through reflection on the life of the Trinity and the revelation that Jesus is the Word, Christianity came to have confidence in the linguistic constitution of the world (Augustine, *On the Trinity*, Shedd edition, 2013, esp. 15.10.17–15.16.26). The Mass contains and gives shape to this reality. Creation really speaks and we can truly understand it. Such understanding is rational, and for most of our common history this ability to listen, to attend, and to hear was not conceived as rationalistic, since it was simultaneously understood to be grounded in the relational life of God, into which we have been created to participate. Such a destiny encompasses all aspects of our being, including our rationality.

Of its very nature, functioning properly, reason appreciates its own limits and bows before God. To have confidence in our capacities as rational creatures is to have a vision of wholeness and unity. Such unity marks us as joint sharers in a common human nature and as the microcosm of the whole of the created order.[14] We live suspended between the poles of brute matter and angelic immateriality, a condition which marks the meaning of rationality.

We may differ, collectively, as regards our opinions on the degree to which we have or have not ever realized our capacities as such creatures. Yet ours is not a crisis of competing systems of thought but one of vision and belief. We need confidence in reason itself, a vision of its capacity to sustain us in prudential common life. For some Christians, the call to prudence sounds from the outset like a concession to situational ethics and non-transcendent horse-trading. Yet a proper understanding of the nature of rationality as part of our species-being shields us from such conclusions and in fact bars the way. Christians, as much as anyone in our times, are called to re-evaluate their conception of reason and to deepen their commitment to it. Our capacity to care for our environment depends upon it. So too does our ability to live and thrive in the world, to be at harmony with it, to enjoy it, and take delight in it.

(4) Beauty beckons. In the urgency of our cultural politics and unrelenting materialism, ordinary people take the lead in affirming beauty, in expressing joy and delight over it. Though the objects of our pleasure can so easily descend to the crass and the vulgar, we constantly affirm beauty's importance in the world. C.S. Lewis once said that "we want something . . . which can hardly be put into words—to be united with the beauty we see, to pass into it, to receive it into ourselves, to bathe in it, to become part of it" (Lewis, "The Weight of Glory," 1980, 16). Beauty both reminds us of our longing for such union and intensifies it. It is integral to the Catholic experience of the Mass, intimately bound up with our common experience of the dependency of nature upon grace and our yearning for self-transcendence.

Thomas Aquinas describes beauty as that which pleases when seen. His saying has been interpreted as urging an attentiveness to what is before us such that hidden dimensions emerge.[15] The beauty of creation and of humanly created objects alike requires our active engagement in order to draw out the moreness that is in them and in the relatedness we share with our environment in our common participation in the divine reality. Beauty bodies forth, it shines from an internal resource of plenitude that is a belonging-to.

However much classical notions of beauty confirmed the importance of this category to our self-understanding, they could never countenance suffering service. It fell to Christianity to prise open the revelatory beauty of self-sacrifice and the inner infinitude of abnegation. Hans Urs von Balthasar points to Augustine's pastoral ministry as the source of a new aesthetic.[16] He avers that if there was no beauty in the One who was slain, this is to be understood not as strict denial but as the recovery of an originary wholeness-in-relation to which we can draw near through the mystery of the cross. Christians reflect this beauty in postures of attentiveness and receptivity such as prayer, in disciplined silencing of the self, and in empathy. The iconoclasm controversy of the eighth century left the Church to conclude that it could no longer be squeamish about images because of the old covenant injunction against them. The incarnation forced the Church to rethink its inheritance and the typological fulfilment of the same in Christ. The Church came to see that it released the arts.[17]

Our natural world inspires thoughts of beauty in us. It elicits responses of praise and worship, from "How Great Thou Art" to "O Holy Night." The Church's history of response can never be reduced to sentimentality or nostalgia, however vulnerable to impoverished reflection re-presentation may persistently be. If Christians are genuinely going to protect the environment, they cannot do so through earnestness, practicality, or even reverence, if beauty is somehow relegated to a position of secondary importance. Beauty is vulnerability, the admission of incompleteness and of longing. It is necessary to our full and proper experience of the world and dominion over it.

"All the Diamonds" and Instrumental Reasoning

I wish to transition now to a discussion of two works that represent areas of Professor Deborah C. Bowen's many interests, popular culture and Canadian literature, in connection with environmentalism, which is of course another of her main areas of interest and scholarly activity. Such a move does not involve a simple shift from theory to practice, or from theology to literature. Sacramentalism resists sharp divisions. "All the Diamonds" by Bruce Cockburn and *For Those Who Hunt the Wounded Down* by David Adams Richards illustrate resistance to instrumental reasoning and political posturing. They achieve effects that discursive argumentation cannot. The sacramentalism practised by these two outstanding artists—one in a pivotal early song, the other in an important early novel—encourages attitudes foundational to a sustainable environmentalism.

Instrumental reasoning implies an approach to things that presumes to restrict their importance and even their reality to our needs and purposes. In my own personal experience, this seems to happen most commonly with reference to economics. Someone will say something on the assumption that our core motivation is self-interest, or they will assume a zero-sum game to explain the world in terms of choice and "opportunity cost." It can be difficult to enter into such discussions, which can occur even among Christians. It seems impossible to address certain assumptions (the conversation seems always already to have taken a particular shape or tone) which do not even have the visibility of assumptions accepted for the sake of argument. Premises, though, are like Trojan horses. They may be invisible, but they're big.

In situations like this, I sometimes find solace, though not exactly a counter-*argument*, in Cockburn's 1974 song "All the Diamonds." In describing the effect of one song, I realize that in some ways I am simply restating the case for the benefit of poetry and art in general, which has the amazing capacity to lift us out of entrenched ways of thinking and cause us to see things differently, even things seemingly completely unrelated to the subject matter of the work of art itself. Nonetheless, I restrict myself to my experience of one particular song and I make the case for its helpfulness.[18]

From the opening line, in which Cockburn's "diamonds" are "conjured," created magically through wind and sunlight on the sea, Cockburn with gentle irony presents an alternative set of values to those of materialism. Real diamonds in this context stand metonymically for whatever material possessions one might covet. The diamonds that mean anything to Cockburn are conjured up by the beauty of creation. Importantly, he finds them beyond nature. At this early point in the song, aided by his soothing voice, one readily imagines an inner harmony and lifestyle of simplicity, contentment, gentleness, and contemplation. Nature, though, provides the entry point. The song begins to lay down the pattern for a rich and holistic environmentalism.

In the years to come, Cockburn will up the metonymic ante, identifying corporate lust for oil and other equivalents of actual diamonds, pointing out government complicity in selfishness, and calling ordinary people to awareness and active resistance: of greed, of political malfeasance, of environmental depredation. In the eighties, he will sing against "a billion burgers worth of beef"[19] grown in deforested areas of the Amazon and on behalf of "this blue-green ball in black space/filled with beauty even now/battered and abused, and lovely."[20] Cockburn's intensifying environmentalism accompanies, or rather attests, a deepening spiritual, Christian attitude towards created reality. His celebrated anti-right-wing politics also make sense as part of a sacramental resistance of an instrumentalizing approach to reality. The politics and the sacramentalism reveal one another.

Cockburn refers to "All the Diamonds" as the turning point in his willingness to identify himself with Christianity in his songwriting.[21] The song includes fairly direct references to the Christian narrative. In it, he makes this turn towards the end of the second stanza, where he is rescued from being "[run] aground" by a ship sent by God that returns him to the open sea. "Thank God" sounds enough like the glib cultural verbal tic to lend the phrase the cover of ordinariness even as it conveys genuine thankfulness and announces the vulnerability of prayerfulness. The chorus's crooned line, "Two thousand years and half a world away/Dying trees grow greener when we pray," gives the song Christian bearings in time and space but does so still somewhat obliquely. Over the years, many Cockburn fans have found this practice attractive—culturally sensitive poetic riddling but declarative nonetheless.

The single word "some" may contain the song's deepest theology: "Thank God he sent some gull-chased ship/To carry me to sea." The randomness and nondescriptness of this "some" folds the whole world of God- and human-made things into larger divine purposes. The Creator might appear anywhere in God-only-knows-what contraption. The whole of reality comes saturated with the possibility of fuller realization, the revelation of God's own active presence.[22] A contrast reinforces this point. One Christian cover of this song changes the word "some" to "his": "Thank God he sent his gull-chased ship. . . ."[23] The change leaves us with a stark difference in showing as opposed to telling. The cover makes explicit the fact that the "gull-chased ship" belongs to God and is an instrument of his providential care. It implies, ever-so-slightly, that other gull-chased ships have a (non-existent) neutral being. Cockburn does not take that tack. In Cockburn's world, all the gull-chased ships in the world stand at the ready, agents of divine service simply by virtue of participating in being, signs of a reckless and profligate divine intentionality.

What still fascinates me about this song, though, is how difficult it is to memorize the lyrics of the last stanza. They confront the listener with a super-saturated abundance barely to be made sense of or contained in the mind as one is led, line by line, from rapture to rapture. Taken together, they overwhelm me, inducing reminiscences of walking up the grade between reedy sloughs as a farmboy going home from school and finding spontaneous whirlpools forming in the spring run-off. The "sea of liquid jade" suggests the image of a nearly infinite rich molten density delicately bearing up the pearl in its compact round smoothness and luminosity. The whole stanza is caught up lyrically in alliteration, liquidity, and exoticism. "Fade" alliterates with "flash" and rhymes with "jade" to intimate fragile wornness and weakness while at the same time attesting hidden depths and inexhaustible resources in darkness as well as light. Listening to the song on my commute on the 401 could lead me to imagine swimming, or is it flying, with a crystal swan in a sky of suns, and leave me speeding.

Beauty, plenitude, yearning, and gratitude expressed poetically in this song for many months overwhelmed argument rehearsals or debriefings in my head against the instrumental economic construal of reality I could never quite vanquish in debate. Even still, when I listen to a song like this one, I am increasingly content to invite my interlocutor into such an experience. I do not say that I break out into song. Rather, the hardheadedness of one sort of reasoning process gives way to a differently figured presentation of a way of being that has an inner logic—a sacramental logic—of its own, one that requires a contemplative measure of time to unfold. Cockburn has followed mystics of the *via negativa* like St John of the Cross and the author of *The Cloud of Unknowing*.

The more sacramental one becomes in one's poeticizing and world-building, the more free the elements in it appear to be and really are. Rowan Williams has written that the world-making of the sacramental artist is "proposed for acceptance and understanding but unable to compel them, since compulsion would make it impossible for the creator to appear as the creator of freedom" (Rowan Williams, *Dostoevsky*, 2007, 12). Instrumental reasoning involves the reduction not just of complexity but of givenness with the end goal of compelling. Part of Cockburn's attractiveness lies in his proposing rather than compelling. The human creator imitates the divine one. The listener or the reader in his or her turn enters into that world thus proposed to find there their own insights and vitality.

Genuine environmentalism must walk this mystical line too. The environment matters for its own sake, and its being is bound up with dependency upon God and God's ordering of creation. The environmentalist—who is, at some level, Everyman—must play the role of creator. She or he neither leaves the created order to its own devices nor forces it into any shape of their own volition. Rather, the environmentalist must persist in proposing—not only in platforms but in enacted schemes—a world liberated from the "harbour towns" of instrumental reasoning.

Coda: It is odd in the extreme that Cockburn should have been stymied for a time by biblical higher textual criticism.[24] His outlook has always tended far more to the sacramental than to the instrumentality such criticism assumes. His delight in story, in community, in retellings, and in revealing hidden patterns finds an analogue in typological reading. This is reading of the sort where the story of Abraham and Isaac is a type of God's sacrifice of his only Son, or where Eve is a type of Mary, who with her "Fiat mihi!" achieves and models what Eve could not. Cockburn returns to this way of poesis in *Bone on Bone*. On tracks such as "Forty Years in the Wilderness" and the Marian "Stab At Matter," he finds his way back to landscape and to matter and celebrates their bearing witness to plenitude.

I was at the same concert as Deborah C. Bowen and her husband John where Cockburn introduced the songs of this album. He performed with energy, joy, and gratitude. The audience were delighted and thankful in return. It is always a pleasure to find people with whom one can share something of the mysteries of art, all the more so when together we can bring to bear Christian modes of exploration and find ourselves illuminating our common faith to one another.

Political Posturing in *For Those Who Hunt the Wounded Down*

David Adams Richards's style also strains sense, though not often with the sparkling beauty of lyrics like "All the Diamonds." It gives up a hard-won beauty. Yet the difficulty of Richards's style opens up worlds. Difficulty characterizes the worlds of his novels, an instance of their definite moreness. In Richards's case, the struggles of life produce a tenuous logic that bespeaks difficult revelation. Such an achievement cannot but challenge glib politicization. If Cockburn reminds one that environmentalism must always ask what it looks like to avoid instrumentalization, Richards warns that it too can be manipulated and that the poor and the downtrodden of society are just as likely to be overlooked by this ism as by any other. For him, posturing and manipulation amount to a failed project and inauthentic form of something intrinsically important.

The opening lines of the New Brunswick novelist's *For Those Who Hunt the Wounded Down* exude difficulty:

> In 1981, during the interval between AA speakers at the meeting in the prison, a man was led in late. He was placed at the back of the crowded room.
>
> It was an AA meeting that Joe Walsh chaired every week here because he felt he was duty-bound to do it.
>
> But the man they had brought in that evening, who sat looking around cautiously, and at the same time impervious to other prisoners, was Joe's nephew, Jerry Bines.
>
> So after the meeting, feeling compelled that he should speak, Joe walked over to him.[25]

On the page, this opening looks like a series of couplets, and it has the uneasy energy of enjambed poetry. The initial, two-sentence paragraph reads simply enough, offering a garden-variety ambiguity by withholding the character's name. We become mildly interested in that character and their story. Taken on its own, the next (one-sentence) paragraph is tolerable too. It gives a character's name, Joe Walsh, and, especially for those who have not read the other books in the trilogy (as well as for those who have), it also poses a question: not now who is this who is being led in late, but why did Joe Walsh feel duty-bound?

Tedium creeps into the third paragraph as Richards's spell begins to take effect. What does the coordinating conjunction "But" coordinate? This is not merely an issue of an author breaking the fussy grammarians' rule of not beginning a sentence (let alone a paragraph) with "But." The issue is that the "But" stands in uncertain and unresolved relation to what has gone before. Whose story does it modify? Does it offer an answer to the question of why Joe Walsh answered the call of duty? Not necessarily. Perhaps the next paragraph will clarify the situation: "So after the meeting, feeling compelled . . ." it begins. We begin to get the sense that Joe feels the weight of duty or compulsion with some regularity, but the "But" still lacks any clear relation to its antecedents. In fact, the "So" deepens the sense of mystery in the connections between the first two paragraphs. We do not know whose perspective we are being invited to take, and we do not know how the opening's inner logic is working, and therefore the logic of the world into which we have just entered.

This slippage marks Richards's prose generally. It indicates class, especially class defined as distance from the bureaucratic and academic centre. Richards's main characters are good, bad, and ugly. Generally, though, they have in common that they work in menial or blue-collar jobs, if they have steady employment at all. They are not typically well-spoken; they often do not have facility with abstract ideas. Yet they experience strange encounters, often involving church settings, and their lives reveal hidden depths, such as the courage to make amends.

Some people in Richards's novels, though, do talk well and think abstractly—professors, administrators, government social workers—but they tend to be his antagonists. The character Vera Pillar belongs to this group. Vera, whose name means "true," works for "the department of social services" (22) and is a writer (23). In the novel, she completes a book about Jerry Bines, a vulnerable man with a sad story who becomes smitten by her for the interest she seems to take in him. She appropriates the telling of that story and makes it conform to a paradigm of patriarchy and male violence that she imposes.

Richards enlists environmentalism as a movement with growing social and political influence in a small but important supporting role in this 1993 novel to illuminate Vera's selfish and highly political agenda. The New Brunswick author associates environmentalism with self-interested political posturing as well as instrumental logic. Paradoxically, it leads out of the world of the novel rather than deeper into it.

Towards the end of the novel, politicized environmentalism emerges randomly as a cause that Vera seizes upon for her own ends. Through his difficult style, he gestures to her interest in the movement with adept clumsiness. Though the story is well advanced by this time, he brusquely introduces her interest in "'important topics' such as the environment" (172). The timing undercuts the importance of the reference. So too does the way he refers to it. He implies that Vera simply asserts the importance of what interests her, and the environment serves only as an example. He presents her as being at least twice-removed from anything like demonstrable environmental concern. A healthier, more sustainable environmentalism would couple genuine concern for one's fellow human being and the rest of the created order as well. Such a mutually reinforcing economy tends to produce more than one anticipates. The context in which Vera operates strongly suggests that the importance of the environment to her is only a politicized one.

With the story nearing its climax, Bines realizes that she has merely manipulated him, showing interest in his difficult circumstances only for the sake of a story that fit her agenda: "Suddenly he realized he was being used for something much more complex than he ever realized" (183). Most importantly, he recognizes that "he was being used," though the acknowledgement of "something . . . complex" suggests Bines's lingering deference to this woman. At the point of this revelation, he has bought Vera a Christmas present, because he has feelings for her, but she is making it clear to him that she is involved with someone else. Richards has Vera gesture to the environment in this exchange with a detail meaningless except as a badge of political correctness and as a means of further wounding Bines:

> "I mean I have a friend—he's studying Environmental Marine Biology in Halifax. Hadley and I are moving down in August if all goes well. He's second-generation Chinese Canadian—so he certainly knows about stereotyping."
>
> She said this to prove that she knew other men with greater problems than Jerry Bines, problems that could in fact involve the court of world conscience. (184)

Vera lives a politicized life, a life that appeals to "the court of world conscience." Someone who works in an environmental area fits what she needs. For Richards, the topic of environmentalism is susceptible to manipulation by educated mouthpieces of government officialdom like her. The way he trivializes the subject, introducing it in a random way and handling it with deliberate clumsiness, might induce some readers to overgeneralize his disdain. Richards, though, takes very specific aim at the way certain individuals, not least those in academic or official government positions, abuse movements like environmentalism for selfish ends.

Vera's brother, Ralphie, works on groundwater research, but at the end of the novel he gives a talk that is attended by only four people, sees a former professor who only pretends to remember him, and in the end decides not to go back to do a Master's: "And he never spoke of it again" (224). The subject of environmentalism simply peters out for the brother. Richards associates it with university life, and the institution of the university is itself a highly politicized milieu, a swamp of vain ambition and pretence. He portrays environmentalism as a cause disconnected from the environment as an extension of his characters' being in the world, the good and the bad characters alike.

The revelation of the meaning of the book's title occurs at the same time that Vera refers to her environmentalist credentials and uses them against the hapless Jerry Bines. After the plot's resolution, her sister-in-law sees the book in a store: "She saw Vera's book on a shelf near the door and glanced at it, but she never bought it. Something about it made her think of it as wounding someone in the heart, hunting someone who was wounded down" (224). Despite their facility with language and bureaucracy, characters like Vera tend to close down opportunities for human connection and human flourishing rather than opening them up.

Vera Pillar's ex-husband, Nevin White, is another more minor character in the novel who ultimately radiates redemptive possibility and gives a hint of what "environment" means to Richards. Nevin is redeemed in a two-stage process. First, he journeys to Fredericton to find his first wife (not Vera) and apologizes to her in one of the intensely moving scenes that populate Richards's novels. Nevin then later finds himself at a local church, helping to raise a cross for Christmas. He asks whether the church ministers could help him after all he had done, and is told, "Of course" (216). Richards then writes, "And he was startled, like Nevin always was, when confronted with simplicity or action" (216). This encounter involves the estranged wife and child of Bines, who is destroyed in the novel.

For Richards, the Church offers both a simplicity and a purposeful action in its liturgical life that replaces bureaucracy and false politics with unforeseeable wholeness and new life. A small church like the one Nevin finds represents the environment that most interests Richards: creation's orientation towards God as revealed pre-eminently in the struggle of people within it. The author makes it difficult to imagine the environment as in any way separable from the stories of people. The movements of grace within the lives of individual persons resonate within particular physical and geographical contexts: they help to render those places more concrete, more real. The grace extended to broken people revivifies their world at the same time. Characters reveal the presence of God in everything through their random encounters with him; in their turn they effect the redemption of the world in their actions.

Richards is no one's idea of a safe pair of hands when it comes to hot-button topics like the environment. He writes with blunt ferocity. He has the great strength of exposing feigned concern that masks self-interest. He recognizes that political posturing damages ordinary people. In such circumstances, environmentalism turns into the demonic inversion of what it could be. Instead of sheltering and nurturing life, it leads to separation and thwarted lives. Political posturing is a form of hunting the wounded down. Yet Richards never simply leaves readers trapped. He affirms hope. It gives his novels substance as well as grit.

For Richards, the environment is the whole world, a world marked by the presence of God. Simplicity of truthfulness and integrity of action reveal that presence everywhere, if the Miramichi region can be allowed to be everywhere. It makes sense that Richards would opt for a writerly habit, like Faulkner's, of setting his stories in a small and seemingly limited geographical context. His *oeuvre* suggests that one can find infinite depths within the givenness of things, and the smallness of the region only serves to accentuate this fact.

In the environment that is the world of any of his novels, an unflattering narratorial style, rough speech, logically strained transitions, and dishevelled narratives challenge the norms of academe and bureaucracy, not to mention those of high culture. It might be appropriate to call a writing style that bears these features realistic, but Richards reveals a presence not easily described discursively. Jacques Maritain describes such realism in *Art and Scholasticism* as the art of reshaping the data of the world to make their structure and relation visible (Maritain, *Art and Scholasticism*, 1930, 90; Williams, *Grace and Necessity*, 2005, 17–18). This is the environment that interests Richards: a world oriented towards God awaiting the revelation of divine presence in action that brings characters to an encounter with their own limitations and encourages them to take some sort of step beyond that. Genuine environmentalism for him begins in the affirmation of such a world and the writing of it.

Sacramentalism persistently challenges one to subordinate human perceptions and orderings of human responsibilities to the prior recognition of the dependency of all things on the Creator. When this recognition soaks our understanding and our loves, which can only happen in a sustaining community, one comes to realize that environmental efforts must likewise flow from the very deepest affirmation of community and of the presence of divine authority in our midst. In the Mass, Christians experience this presence and this community, even as it urges all who participate in it to affirm the life of the world and to seek its flourishing. Stewarding the sacred mysteries, the Church stands with arms open to the world, inviting it into its upward journey, sharing its pain, listening and learning from it, celebrating what it knows, revealing and guarding the uncommonness of our common home.

Notes

1 For a gentle, poetic description of the Catholic understanding of all the elements of the Mass, see Romano Guardini, *Sacred Signs*, 1956.
2 For a very good introduction to the theology of participation, see Boersma, *Heavenly Participation*, 2011, esp. 1–51.
3 Very good on describing the attitude to the liturgy that Catholics and Orthodox share is Schmemann, *For the Life of the World*, 1973. On the point of re-ordering, see especially 26–28.
4 "Instrumental" has two distinct meanings. When I treat something or someone merely as an object for my own purposes in the moment (or for any length of time in a non-transcendent way), I tacitly claim to understand that thing or purpose in a totalizing way, which is absurd since their reality derives from their participation in God's reality. I use them in a sense that cannot but be idolatrous. Augustine teaches the importance of using in a very different sense. In the proper ordering of loves, the *ordo amoris*, I am only to seek to enjoy God. Everything (and everyone) else is to be subordinated to the desire for God and the enjoyment of him. Everything else is, relatively speaking, to be used rather than enjoyed.
5 It matters a great deal that the Church has an unbroken line of sacramental representatives going back to its founding. The continuity enacts ontological participation and has helped preserve the Church against the threat of gnostic claims to secret learning.
6 Aaron Riches offers a beautiful recent meditation on this reality in "Sacrifice and Freedom," 2018, especially 77–84.
7 For an introduction to this very important topic, see Cunningham, *Natura Pura*, 2010.
8 On the idea of creation as going out from and returning to God, the *exitus-reditus* scheme, see Ratzinger, *The Spirit of the Liturgy*, 2000, 29–34.
9 N.T. Wright is particularly good on the covenantal role that the people of God play in the divine economy. See *Surprised by Hope*, 2008, 189–295, among his many writings on this theme.
10 For a particularly rhapsodic and energetic description of the theology involved here, see Kruger, *The Great Dance*, 2000, 57–65.

11 On the problem of emotivism, see especially MacIntyre, *After Virtue*, 1985, 6–35.
12 I explain both the necessity of the development of the critique of rationalism and its limitations in the forms known as literary, cultural, or critical theory in *Rationality Is . . . The Essence of Literary Theory*, 2022.
13 Against the modern understanding of self-interest, in a view indebted to Platonic ontology, see Grant, *English-Speaking Justice*, 1998, 44–47. For a similar critique specifically on Machievelli's doctrine, see Maritain, "The End of Machiavellianism," 135–163.
14 The early modern thinker Nicholas of Cusa offers a very strong description of man as microcosm in *On Learned Ignorance,* 3.3.195–198.
15 For a classic modern discussion of Thomas's theory, see Maritain, *Art and Scholasticism*, 1930. For a contemporary discussion of Maritain's aesthetic theory, see Williams, *Grace and Necessity*, 2005, 1–42.
16 Balthasar, "Revelation," 144–155. See also David Bentley Hart, *The Beauty of the Infinite*, 2003, 122–125.
17 Klassen, "Beauty is the Church's Unity," 2020, 63–76.
18 Bruce Cockburn, "All the Diamonds" (1974), *Circles in the Stream*, Toronto: True North Records, 1977. Full lyrics can be read at http://cockburnproject.net/songs&music/atd.html.
19 Bruce Cockburn, "If a Tree Falls," *Big Circumstance*, Toronto: True North, 1988.
20 Bruce Cockburn, "Planet of the Clowns," *The Trouble with Normal*, Toronto: True North, 1983.
21 "'I became a Christian in 1974, officially, to myself,' he explained. 'Before that, I was kind of heading that way, but I didn't perceive it as a real personal involvement requiring any kind of commitment until after a certain point in the summer of '74, and then I started writing about it. The song "All The Diamonds In The World" was the song that sort of marked that turning point'" http://cockburnproject.net/issues/personal/christianity.html (3 April 1992).
22 Cockburn's image of a ship carrying him to sea may appear to offer a romantic picture of liberty, but the lyricist does not suggest that God always leads one away from people or involvement in local affairs. Romanticism *can* result in weak politics, but for Cockburn the freedom of belief orients him towards the world. The evidence of his unfurled career as a songwriter and activist indicates his involvement. He sees all of life as full of depth and mystery, a sea of adventure—as much communal as personal if one is taking delight in God and responding to him.
23 I'm going to follow C.S. Lewis's lead from *The Abolition of Man* here and preserve the anonymity of a source.
24 This attitude is hinted at on the opening page of his autobiography. Cockburn, *Rumours of Glory*, 2015, 1.
25 David Adams Richards, *For Those Who Hunt the Wounded Down* (1993), Toronto: McClelland and Stewart, 2000, 1. Subsequent page references are given in the text.

References

Augustine of Hippo, Saint. *On the Trinity—St. Augustine of Hippo*. Translated by Arthur West Haddam. Edited by Paul A. Boër. Revised, annotated, and introduced by William G. T. Shedd. Houston, TX: Veritas Splendor Publications, 2013. Kindle edition.

Boersma, Hans. *Heavenly Participation*. Grand Rapids: Eerdmans, 2011.

Cockburn, Bruce. "All the Diamonds." 1974. In *Circles in the Stream*. Toronto: True North Records, 1977.

———. "If a Tree Falls." In *Big Circumstance*. Toronto: True North Records, 1988.

———. "Issues: Personal: Spirituality/Christianity." *Archive of Comments to Various Interviewers*. http://cockburnproject.net/issues/personal/christianity.html.

———. "Planet of the Clowns." In *The Trouble with Normal*. Toronto: True North Records, 1983.

———. *Rumours of Glory: A Memoir*. Toronto: HarperCollins, 2015.
Cunningham, Conor. "*Natura Pura*, an Invention of the Anti-Christ: A Week with No Sabbath." *Communio* 37 (2010): 243–254.
Dupré, Louis. *Passage to Modernity*. New Haven: Yale University Press, 1993.
Grant, George. *English-Speaking Justice*. Toronto: Anansi, 1998.
Guardini, Romano. *Sacred Signs*. Translated by Grace Branham. St. Louis, MO: Pio Decimo Press, 1956. www.ewtn.com/catholicism/library/sacred-signs-11190.
Hart, David Bentley. *The Beauty of the Infinite: The Aesthetics of Christian Truth*. Grand Rapids: Eerdmans, 2003.
Klassen, Norm. "Beauty Is the Church's Unity: Supernatural Finality, Aesthetics, and Catholic—Orthodox Dialogue." *Analogia: The Pemptousia Journal for Theological Studies* 10 (2020): 63–76.
———. *Rationality Is . . . The Essence of Literary Theory*. Eugene: Cascade, 2022.
Kruger, C. Baxter. *The Great Dance: The Christian Vision Revisited*. Jackson: Perichoresis, 2000.
Lewis, C. S. *The Weight of Glory and Other Addresses*. New York: Macmillan, 1980.
MacIntyre, Alasdair. *After Virtue*. 2nd ed. London: Duckworth, 1985.
Maritain, Jacques. *Art and Scholasticism*. London: Sheed and Ward, 1930.
———. "The End of Machiavellianism." In *The Range of Reason*, 135–163. London: Geoffrey Bles, 1953.
Nicholas of Cusa. "On Learned Ignorance." In Nicholas of Cusa, *Selected Spiritual Writings*. Translated by H. Lawrence Bond, 85–206. New York: Paulist Press, 1997.
Ratzinger, Joseph. *The Spirit of the Liturgy*. Translated by John Saward. San Francisco: Ignatius, 2000.
Richards, David Adams. *For Those Who Hunt the Wounded Down*. Toronto: McClelland and Stewart, 1993, 2000. All quotations are from the 2000 edition.
Riches, Aaron. "Sacrifice and Freedom: On Mystagogy and the Way of Atonement." *Communio* 45 (2018): 54–84.
Schmemann, Alexander. *For the Life of the World*. Crestwood: St. Vladimir's, 1973.
Von Balthasar, Hans Urs. "Revelation and the Beautiful." In *Word and Revelation: Essays in Theology I*. Berlin: Herder and Herder, 1964.
Williams, Rowan. *Dostoevsky: Language, Faith, and Fiction*. Waco: Baylor University Press, 2007.
———. *Grace and Necessity*. London: Continuum, 2005.
Wright, N. T. *Surprised by Hope*. New York: HarperOne, 2008.

14 Poems by John Terpstra (II)

Giants

There used to be giants,
and they loved it here. They'd sit
their giant hinds in a row along the top edge
of the escarpment, and pick at the loose rock
with their hands or their feet, then throw or skip
the smoothest stones across the bay, to see who could land one
on the sandstrip, three miles away;

or they'd spring themselves off the scarp top
like you would off a low wall, and go running
all the way to the end of the sandbar,
and jump across the water to the other side,
or jump in, splashing and yelling up the ravines,
chasing each other's echoes.

This was only a few thousand years ago,
and the giants were still excited about the glaciers,
which were just leaving; about not having to wear
their coats all the time, and what
the ice and water had done, shaping and carving
this gentle, wild landscape!

They loved it here.

I'm telling you, they absolutely loved
every living minute here,

and they regretted ever having to leave.

* Previously published in *Disarmament* (Gaspereau Press, 2003), and reprinted here by kind permission of Gaspereau Press.

DOI: 10.4324/9781003366744-14

A Brief History of Settlement on Hamilton Mountain, From 1789

We came to a place
after many days walking,
the top of a cliff,
a plateau of wood and meadow,
and followed a creek
to its sweet water's source
below an outcrop of rock
tall enough to provide a broad view
of this oasis from the strife we had so recently suffered
in another land.
We gave thanks, and chose to stay,
as did all our company, each family planting itself
beside spring and pond
of creeks whose paths they traced
from waterfalls over that cliff face, the scarp
which served as a protective rampart to our community
for seven generations,
until the young city mounted its height,
stormed our fields and woodlots,
buried creeks in pipes,
and sent springs into hiding
beneath the trees,
the few left standing—
                                        the trees,
those intermediaries between ourselves
and our Maker,
whose roots never cease to divine the living waters
that my people, naked
under shower heads, with kitchen faucets
running full blast,
                                forget.

The Once and Future Creek

Then the angel showed me the river of the water of life,
bright as crystal . . . , flowing through the middle of the street

Revelation 22

A surprise creek runs bright
between cracked and weedy slabs
of concrete, through a vacant lot
at the bottom of Queen Street,
where railway tracks block the shoreline of the bay.

The creek flows past piles of brick
and a row of container boxes filled with
god-knows-what,
slips under a chain link fence
and crosses the sidewalk, falling
over the curb and through a storm drain grate.
It has flowed for long enough now
that marsh grasses line its edge,
and two ducks swam away upstream, hurriedly,
looking over their shoulders,
when I came by to take photographs.

The creek runs like a long lost memory
through a desolate and redundant
industrial area,
the first landscape of the city
to be forested in smokestacks,
seven generations back,
where a century-old, bird's-eye map
shows ravines, and an older map
their watercourses, one of which, impossibly,
matches the path of this clear blue line.

If the deal doesn't fall through,
this vacant lot will become home
to a movie studio,
a big box
in which imaginary worlds will be created,
while the real world
        is blocked out and buried,
a world to which,
thanks to a daughter of the city

who pointed me to this rill, this trickle,
I am even more
        perilously attached.

Seven generations. It's so unkind.
What can I tell the children? I'm sick
of finding reasons and making excuses
for the naked disrespect,
most of which have to do with money.

Although I dare not lap or dip a cup,
I take this creek, whose source
is likely a left-on tap or leaking water main,
yet may just be a hidden spring,
as revelation.
        I read this brooklet
as a line of hope
against hope,
        speaking from the future.

15 Birding, Fiction, and Margaret Atwood's Cultivation of Ecological Awareness

Tina Trigg

As an internationally recognized Canadian author and activist, Margaret Atwood's influence on popularizing representations of environmental issues is strongly associated with an apocalyptic vision. This expectation culminates in her dystopic trilogy of speculative fiction—*Oryx and Crake* (2003), *The Year of the Flood* (2009), and *MaddAddam* (2013)—with their unflinching portrayal of human-engineered environmental disaster and the painstaking, precarious ecological recovery that ensues. The series has had considerable critical attention, including from Christian scholars like Deborah C. Bowen, who argues that not only do "these novels articulate and critique the cultural and ecological sensibilities of our day" but that Atwood's "apocalyptic vision resonates rather surprisingly with the ending of what Christians understand to be the ultimate story of eschatological hope: the vision in the Book of Revelation . . . [of] 'a new heaven and new earth'" (Bowen, "Ecological Endings," 2017, 692). Bowen carefully outlines the open endings of Atwood's trilogy as aligning with Paul S. Fiddes's conception of "a truly religious view of the end of all things": a balance of "certainty about the overcoming of evil and the triumph of God's purposes" with necessary latitude for a limitless God and for "the freedom of human beings to contribute to God's project in creation." This balance of certainty and uncertainty cannot then be prescriptive but rather "demands an openness in the future" (Fiddes, *The Promised End*, qtd in Bowen, 2017, 698). And, importantly, this future—both Biblical and in Atwood's trilogy—is carefully situated in a restored physical earth, not in a vague, nebulous "better place" or outer space but here, renewed. As Atwood affirms, the novels form

> a series exploring . . . our own planet in the future. (I carefully say *a* future rather than *the* future because the future is an unknown: from the moment *now*, an infinite number of roads lead away to 'the future,' each heading in a different direction).
>
> (Atwood, *In Other Worlds*, 2011, 5)

Hence, Atwood terms her *MaddAddam* novels "speculative fiction" not science fiction, for they encapsulate "things that really could happen but just hadn't completely happened when the authors wrote the books" (Atwood, *In Other Worlds*, 2011, 6). While tracing a possible trajectory based on real-world observation,

DOI: 10.4324/9781003366744-15

the novels also balance certainty with uncertainty, known events with unknown possibilities.

Atwood contends that tracing these vast possibilities in her speculative fiction is a storytelling response to "questions people are increasingly asking themselves: How badly have we messed up the planet? Can we dig ourselves out? What would a species-wide self-rescue effort look like if played out in actuality?" (Atwood, *In Other Worlds*, 2011, 94). Clearly, her concern is for both possibility and limitations, suggesting an ecological frame of reference; no organism is fully independent and, hence, each is responsible to the others, in the moment and for the future. While frank about the disparity among contemporary Christian communities concerning environmentalism and ecological ethics, Bowen explores Atwood's early linkages with the Canadian branch of interdenominational Christian environmental organization A Rocha and how such partnerships might signal restorative human re-engagement with creation. Begun in 2014, Atwood's connection with A Rocha continues to flourish; in 2020 A Rocha Canada's British Columbia conservation campaign centred on a significant matching gift pledge from Atwood (https://arocha.ca/portfolio-items/atwoodmatch/). Throughout her long career, Atwood has continually insisted on reader-complicity in *creating* the worlds they inhabit—both fictionally and actually—and so on being accountable for their actions or apathy. Similarly, A Rocha's organizational self-description emphasizes future hope through present-day small, sustainable change by ordinary people; it is a call to meaningful action and to making a collective difference: "In an environmental context often characterized by fear and paralysis, A Rocha is bringing hope through care of both people and places. . . . A Rocha is changing the way people treat the earth" (https://arocha.ca/). Complementing environmentalism's focus on place—that is, "the way people treat the earth"—the broader ecological turn in Atwood's writing is suggestive in its emphasis on relationality, interdependency, and adaptation. However, beyond the God's Gardeners cult and its survivors in the *MaddAddam* series, this broader ecological emphasis has not been widely recognized.

Moreover, what is less readily apparent is that, well before the recent dystopic series, these ecological concerns of relationality, interdependency, and adaptation have been quietly present in Atwood's work across several genres and from early writings. In counterpoint to the overshadowing criticism on the *MaddAddam* trilogy, this chapter discusses an early short story ("The Resplendent Quetzal," *Dancing Girls*, 1977) and a recent graphic novel series (*Angel Catbird*, 2016–17) to demonstrate how Atwood's focus on birding cultivates an ecological awareness that moves beyond literary representations to punctuate her activism and to advocate for a biocentric, sacramental understanding of creation, rather than an anthropocentric one. In terms of conservation, of particular note is Atwood's emphasis on intergenerational collective action and youth as sources of possibility, change, and hope for the future.

In "The Resplendent Quetzal" (*Dancing Girls*, 1977) birding is represented as both catalyst and camouflage for the married protagonists, Sarah and Edward. What began as a vulnerable, intimate sharing of self has fizzled into a persistent irritation:

> His passion for birds had been one of the first things Edward had confided to her. Shyly, as if it had been some precious gift, he'd shown her the lined [bird

identification] notebook [of his youth]. . . . She'd pretended to be touched and interested, and in fact she had been.

(Atwood, "Resplendent," 1977, 164)

Sarah's retrospective self-description suggests an ongoing alienation from her emotions and her partner: initial pretence became a genuine but fleeting connection. Despite her simmering resentment, Sarah continues to go birding with Edward though even her feigned interest is fading: "she'd let him think she shared or at least indulged his interests. She was becoming less indulgent as she grew older" (Atwood, "Resplendent," 1977, 165). Sarah's alienation is not only from herself but evidently from Edward and their relationship as well; rather than increasing interdependency and intimacy over time, their married relationship is represented as a shell—increasingly hard, hollow, and self-protective.

Alice Ridout suggests that reverting to familiar (even unfulfilling) routines can be seen as a coping mechanism for Atwood's characters, as each "attempts to escape the pain of linear time and the narratives that unfold through it" (Ridout, "Temporality," 2000, 862). Certainly, birding has become a default habit that provides some consistency to their otherwise tenuous relationship. It is a predictable narrative for them to inhabit. Similarly, Sarah and Edward default to accepted period social roles of husband and wife on a single income: he, a sixth-grade teacher; she, a "housewife." Not only do they make birding a sporadic recreational activity, but, to Sarah's suppressed frustration, their vacations are consistently centred on birding rather than on leisure, comfort, or connection. As such, relocating to far-off seemingly exotic places only seems to increase the emotional distance in their relationship; their counterposed desires in travel (resting versus sightseeing, familiar comfort versus authentic local fare and lodging) suggest a deeper dissatisfaction or relational discontent. As Reingard M. Nischik notes, in this story collection, Atwood often presents characters "separated from their accustomed social surroundings (for example . . . abroad), reinforcing the rootlessness of these 'tourist' characters," that is, characters who distance themselves from others as passive observers (Nischik, "Margaret Atwood's Short Stories," 2006, 148–149). While Edward flits among interests, discarding one for the next, Sarah observes from the sidelines, no longer fascinated but increasingly fatigued by his restlessness. However, Sarah observes that "Edward thumbed through his *Peterson's Field Guide* as if it were the Bible or the bird was the Holy Grail" (Atwood, "Resplendent," 1977, 165). Atwood directly represents Edward's reverence, commitment, and fastidious attention to detail in birding as religious in fervour; conversely, Sarah's participation is perfunctory and superficial. Unlike birding, which Sarah identifies as Edward's only enduring interest, Sarah categorizes herself as passé, a discarded interest among many others: "She herself, she thought, had once been one of his obsessions" (Atwood, "Resplendent," 1977, 165). However, rather than an obsession, birding might also be called an embodiment of his all-consuming desire, the quest around which Edward organizes his life. In James K.A. Smith's terms, Edward is exhibiting what it is to be human, being "on the move, pursuing something" (Smith, *You Are What You Love*, 2016, 8), but he may also be experiencing a misdirected desire. Birding demands far less from him than authentic human relationship would, so he continues to pour himself into it and so to avoid truly knowing Sarah in the sense philosopher Martha Nussbaum describes love's knowledge: as an intimacy that is necessarily vulnerable, deep, trusting, and with an acceptance of incompleteness (Nussbaum, *Love's Knowledge*, 1990, 282).

Significantly, Atwood does not limit this duality of catalyst and camouflage to Sarah; both partners knowingly use birding as a camouflage for relational fractures. What began as a courtesy quickly becomes a defining feature of their relationship: "of course she'd gone [birding]. . . . She hadn't complained, then. . . . She'd even become quite good at it" (Atwood, "Resplendent," 1977, 165). But soon birding becomes a kind of decoy for an animate relationship: Sarah was "quicker at spotting movement than he was . . . [and] she'd fallen into the habit of using it when she wanted to get rid of him for a while" (Atwood, "Resplendent," 1977, 165). While Atwood carefully opens the story from Sarah's perspective, Edward's supposed naïveté is just as quickly overturned when the narrative shifts to his point of view. He fumes about her inability to perceive his quiet rage and inwardly rails against her dismissive contempt. Although Edward acknowledges his complicity in keeping up the charade, he is as baffled by his own motives as by hers: "why did he do it, why did he chase off after her imaginary birds, pretending he believed her," and why was she doing this to him? (Atwood, "Resplendent," 1977, 166–67). Nonetheless, Edward is chillingly aware that he is evading "the real reason" out of fear and so their lives are built on falsehoods and pretence; he believes any confrontation will reduce it to "rubble. . . . There would be nothing left to say and Edward wasn't ready for that" (Atwood, "Resplendent," 1977, 167). The interdependency they have achieved is one of dysfunction wherein birding masks multiple forms of pain and trauma.

In "The Resplendent Quetzal," human relationality and adaptation are, in fact, exposed as responses to loss—including Edward's birding and Sarah's destruction of a Christ child icon. Rather than a culmination of intimacy, Sarah and Edward's relationship is more like an afterthought, aligning with the social conventions of the time for unplanned pregnancy. On the surface, Edward makes the pragmatic choice to leave the study of theology to take a teaching certificate. However, it is arguable that he transfers the holistic conceptualization of creation from a traditional theological frame to an ecological one, particularly as it materializes in birds. That is not to say that he creates an ornithological deity or mythology but rather that birding becomes both an act of worship that characterizes his life and becomes his organizing principle, his centre. He adapts to a diminishment of his world by reconstituting it as primarily material. Yet, even in this framing, Edward largely evades any sense of ecological interdependency by carefully situating himself in the role of observer. His presence or absence is inconsequential to the birds and their way of living; he just watches from a distance.

Beyond Edward's abandonment of theological studies, Atwood hints at the importance of religious ritual, place, and the power of cultural narratives through the sacrificial well. In the opening scene, Sarah is described as "sitting near the edge," then immediately distanced from the sacrificial well as a tourist who is both sceptical of its value and scornful of its very appearance (Atwood, "Resplendent,"

1977, 163). By the end of the story, however, she purposefully "walk[s] slowly towards the well, until she [is] standing on the very brink," where she flings in not herself but a plaster Christ child that she stole from a seedy local restaurant (177). On the one hand, her replacement of the community's past virginal human sacrifice with a benign (and tacky) chipped plaster icon seems to refute both the power of local Aztec myth and Christian conceptualization of the perfect holy Christ child of virgin born—or at least to suggest they are interchangeable mythologies that she equally discards. On the other hand, intentionally participating in the symbolic sacrifice and death of this broken Christ contrasts with all of Sarah's human efforts to sustain or create life; her care during pregnancy comes to naught with her own "perfect child" stillborn in a "freak accident" (Atwood, "Resplendent," 1977, 168). Unable to adapt her narrative to this traumatic event, Sarah withdraws emotionally and tightly draws her boundaries to the domestic sphere. She neither continues her studies nor works outside the home but concentrates her life around Edward. Though intentionally circumscribing herself, Sarah ends up feeling suffocated: "He was omnipresent, he pervaded her life like a kind of smell; it was hard for her to think or act except in reference to him" (Atwood, "Resplendent," 1977, 169). At the same time, she feels resentful about being abandoned: "He had left her alone with the corpse, a corpse for which there was no explanation" (176). Both Sarah and their marriage survive but in a diminished state. There is no relational flourishing here of body, heart, or spirit. Atwood represents human relationality and adaptation as responses to loss—including Sarah's theft and destruction of the Christ child icon, materialized as a complex, contested gesture towards the unseen, unknown, and uncontrollable.

Like the rare titular bird, an ecology of relational flourishing remains elusive but desirable for both characters and readers of this short fiction. Alice Ridout notes that "In *Dancing Girls* Atwood portrays many characters who are prisoners of life scripts which they wish to escape but cannot" (Ridout, "Temporality," 2000, 852). Sarah and Edward in "The Resplendent Quetzal" are no exception. Edward's transcendent awe hinges on birds remaining imaginative possibilities; it evaporates when they are concretely before him. Yet he also acknowledges that their collective inexhaustibility fuels his faithful pursuit: "there were hundreds of kinds he would never see; no matter how many he saw there would always be one more. Perhaps this was why he kept looking" (Atwood, "Resplendent," 1977, 172). There is both a certainty in their presence and an uncertainty in the particulars of time, place, or variety: a balance of known and unknown, an expectancy for the future. In contrast, this perception of possibilities and the metaphorical aspects of birding—flight, escape, freedom—are contained in the narrative through the title's specificity. Sarah's insistence on seeing a particular bird, a Resplendent Quetzal, reveals both her character and her pain. Although the bird does not inhabit the area they are visiting, she is unwavering in her expressed desire: "Well, that's the one I want [to see]. . . . That's the only one I want." Atwood layers the text with connections of loss, birds, spirituality, birth, and unborn children: birds are attributed to being souls of the unborn, the resplendent quetzal's habitat is cloud forests, quetzal means feather, and the Aztecs called unborn babies "A jewel, a precious feather"

(Atwood, "Resplendent," 1977, 173). Sarah is characterized as stubborn, controlling, and unflappable, yet she could neither predict nor prevent their stillborn child. Hence, while Edward still believes—in wavering faith—that another child might arrive, acknowledging "[t]hey came when you least expected it. A jewel, a precious feather," Sarah exerts control to prevent this likelihood by secretly taking contraceptives (Atwood, "Resplendent," 1977, 175). Her fear of recurrent loss manifests as control, deception, and emotional distancing; in dismay she watches her stalled, unsatisfying life instead of living it.

Flinging the nativity child into the sacrificial well is an act that breaks Sarah's defences; however, as her loss finally materializes in tears, Edward no longer recognizes her. He desperately consoles and cajoles but is unable to meet her in the place of unspeakable suffering: "[With] [t]he ordinary Sarah . . . he'd invented ways of coping. But he was unprepared for this. She had always been the one in control" (Atwood, "Resplendent," 1977, 177). Yet there is a significant difference now that Sarah has acknowledged her pain, anger, and lack of control; instead of being alienated, she redefines her identity to include her emotions and the act of loving to include the pain of loss. While Edward acknowledges his fear of adapting to her vulnerability in this final scene, he does not move past it. Instead, he stubbornly clings to their familiar identities and tries to contain them in known, though painfully inadequate, roles. As a result, the story closes with them at silent loggerheads and Sarah's question "Did you find your bird?" still hanging in the air. The opening scene's false quest and pretence became reality: the imaginary bird *actually* appeared and again disappeared without warning; however, the child (the jewel, the precious feather) was irrevocably "lost." Atwood creates narrative certainty and uncertainty related to what the characters think they most desire.

Whether Sarah and Edward's relationship will transform from pretence to relational intimacy may seem unlikely but nonetheless *does* remain one possible outcome among many. Nischik contends that in Atwood's first short story collection, *Dancing Girls*, stories of "partnerships in crisis" are "suffused with desperation and hopelessness" while "a glimmer of hope" characterizes her subsequent story collection, *Bluebeard's Egg* (1983) (Nischik, "Margaret Atwood's Short Stories," 2006, 149). However, I would argue that the open-ended conclusion of "The Resplendent Quetzal" already contains such "a glimmer of hope." Just as trauma can radically disrupt the likeliest trajectories, Atwood insists on human choice and agency as constant possibilities that can redirect expected narratives to other places, better *or* worse than the present. There are no guarantees in relationships and authenticity does entail risk; however, so does alienation even when masking as self-protection. In fact, the latter is far more risky—and yet common. Shame and vulnerability researcher Brené Brown argues that being authentic takes courage and is "a practice—a conscious choice of how we want to live. . . . It's a choice to show up and be real. The choice to be honest. The choice to let our true selves be seen" (Brown, *Gifts*, 2020, 67). In this respect, Atwood's stories are relentless representations of possibility; they typically conclude unresolved, poised in moments when one can choose otherwise. As she reminds us, "the future is an unknown: from the moment *now*, an infinite number of roads lead away to 'the future,' each heading in a different direction" (Atwood, *In Other Worlds*, 2011, 5). Similarly, Brown observes that "[a]uthenticity is a collection of choices we have to make every day"; hence, the possibility for "wholehearted living" is ever before us (Brown, *Gifts*, 2020, 67). That is a "resplendently" hopeful possibility.

As a complement to this early work, Atwood's recent graphic novel series, *Angel Catbird* (originally published 2016–2017), illustrates ecological concerns of relationality, interdependency, and adaptation in literal ways across three volumes and across three species: human, domestic feline, and owl. In contrast to the traditional form of short story in "The Resplendent Quetzal," Atwood's graphic novels are a site of disruption: in content, form, and consequently in reception. Hired and then targeted by a half-rat (human who can change into a rat) who is mobilizing his rat army to take over the world, Atwood's protagonist teams up with half-cats and half-owls to preserve biodiversity. As Jonathan Carey points out:

> For Atwood, *Angel Catbird* is her way of creating common ground between cat lovers and bird lovers, who too frequently see each other as enemies. . . . Her hero finds himself caught between two worlds—feline and avian—as he comes to develop and understand his different personalities.
>
> (Carey, "What's Part Cat," 2016)

Unable to destroy the enemy that is now part of himself, Strig Feleedus is confronted by competing desires, natural human, feline, and bird instincts that are at direct odds. Shortly after the super-splicer formula changes him into half cat-half owl, Strig encounters some cats with a baby bird; while his human instinct is to return the bird to its nest, he finds himself unable to act "on instinct." Instead, he hesitates and wonders, "Do I rescue it or eat it? . . . I want to eat it. But I can't. Something's holding me back. Anyway, I'm not a cat. Or am I?" Chastised by the alley cats, as a "Traitor to cats" for his "Predator fail," Strig acknowledges his error, then asking himself, "What's wrong with me?" (Atwood, *Complete*, 2018, 39–40). The question is ironic since there is no longer a single perspective providing him with high moral ground. Although speaking in human physical form, as a hybrid being, he no longer inhabits a singular dominant place in the food chain; hence, his ecological ethics become tangled and more complex.

Based on comic historian Jeffrey Johnson's assertion that "these early superheroes were products of their time and social avengers that fought for the average man," scholar Esther Susan Csorba contends that super-heroes were "originally constructed with the sole purpose of entertaining the average man in the venture of being able to destroy evil" (Csorba, "Unlikely," 2019). She dismisses Atwood's protagonist from this category, arguing that "Angel Catbird is not a social commentary, nor was [the character] born with the intent of becoming the ultimate salvation" (Csorba, "Unlikely," 2019). I would disagree; this inversion of a single white male hero may be precisely the point that makes Atwood's character a social commentary in ecological terms—to destroy (or stay) the evil of ecological collapse requires sustained collective action, not a single heroic action by one individual. Therefore, logically, Atwood's graphic novels pointedly refuse the narrative of a single saviour. If ecology is characterized by interdependency, then waiting passively to be saved instead of engaging in renewal may be, in fact, to participate in ecological destruction; in Atwood's view, "passive acquiescence does not absolve guilt or remove responsibility" (Grace, *Violent Duality*, 1980, 3). Re-framing the average person's perspective to an ecological one requires diffusing polarization in favour of collaboration, and self-interest to include other-interest across environments and species. This is a form of biocentrism, what philosopher Arne Naess calls "deep ecology" (*Ecology, Community, Lifestyle* qtd in Benshirim, "Daring to Care," 2016). It is important to recognize that Atwood's texts do not presume the ultimate salvation or redemption of creation by a Christian God or any other external force; the representation is humanist. However, by emphasizing interdependency, relationality, and adaptation, the narratives do allow for human participation in restoration of society and the earth—or, in Fiddes's terms "the freedom of human beings to contribute to God's project in creation" (*The Promised End*, qtd in Bowen, "Ecological Endings," 2017, 698). Atwood represents all those human-animal characters fighting to stop the megalomaniacal rat-man, Murdoch, as social avengers. They are collective heroes—a category Atwood identifies as a definitively Canadian trope. In *Survival* (1972), she contends that "[t]he attempt to create a traditional individual hero may be misguided in Canada" not only historically but particularly beyond the twentieth century (Atwood, *Survival*, 1972, 172, emphasis in original). If traditional male heroes "must be central, significant: [with] issues and kingdoms depend[ing] on his rise and fall" (Atwood, *Survival*, 1972, 165), then shifting the heroic role across both species and genders is significant. Creating collective social avengers, rather than an individual one, coheres Atwood's representation of character with *Angel Catbird*'s contemporary biocentric narrative.

Not only is the series based on disruption in the form of human-engineered interspecies modification and a disruption of the singular hero archetype, but Atwood also shifts the focus of these graphic novels away from pure entertainment towards unstinting instruction. Interestingly, the history of graphic novels as a genre had more to do with authors "aim[ing] for a distinction and an emancipation within the comics field—to be able to freely tell stories . . . without having to take into account an audience of children or to limit one's ambitions" (Méon, "Will Eisner," 2017). Perhaps since the genre of graphic novel has been established since the mid-twentieth century, Atwood is now able to do both: *Angel Catbird* tells an ambitious ecological story that contains hope for future change through collective effort yet is targeted at children. While Atwood's emphasis on the hope for "broader social renewal" has been noted by critics for decades (Hengen, *Margaret Atwood's Power*, 1993, 16), the intensity of message and mode of address are distinct in this graphic novel series.

More overtly than her other works, *Angel Catbird* incorporates didactic (at times even moralizing) statistics, tips, and ecologically friendly action steps for readers that parallel the novel's storyboard action, signalling the interplay of literature and our way of being in the world. For example, Count Catula counsels other social avengers (the "good guys") against using the label of "monster" for Murdoch, their foe, since it's "a loaded word [that] gives people ideas." Embedded on the same page is a direct address to pet owners to adopt a nuanced perspective of animals' actions in context and calling them to take personal accountability: "Cats are hunters but they're not monsters. . . . Cat owners are responsible for their pets' behaviour" (Atwood, *Complete*, 2018, 115). The tips are consistently followed by an invitation to visit Nature Canada's website, an organization whose "mission is to protect and conserve nature in Canada by engaging Canadians and by advocating on behalf of nature" (https://catsandbirds.ca/). Offset on white blocks at the bottom of pages filled with colourful panels of stunning illustrations by Johnnie Christmas, these advocacy tips might easily be overlooked, but they are cleverly worded and varying enough in content to appeal to a curious reader. Possibly, the sheer quantity of the advocacy tips also increases the likelihood of readers engaging with at least some of them.

While Atwood's text is appreciated by high visibility organizations like Audubon, even they acknowledge that "the conservation message isn't exactly subtle. The pages are littered with factoids on birds and cats, discussing everything from felines' acute sense of smell to the urgency of keeping pets indoors." While critical, they also draw attention to how publishers and organizations with shared values can partner effectively to highlight our interdependencies: "The first instalment of *Angel Catbird* was released in tandem with Nature Canada's 'Keep Cats Safe and Save Bird Lives' program targeting domestic cat owners to keep cats indoors by raising their awareness of avian precarity in urban centres" (Carey, "What's Part Cat", 2016). In this respect, as Lorraine York observes, Atwood takes "a nuanced and strategic approach to literary production—and the consequent production of literary celebrity"(York, *Margaret Atwood*, 2013, 24). She cultivates cultural capital as a significant form of power which, in this case, Atwood wields for collaboratively raising ecological awareness in the reading public. Savvy readers, York maintains, will acknowledge the economic conditions and political limitations of global literary production but also will recognize (and presumably appreciate) "those instances when the commercial intertwines productively with the artistic" (York, *Margaret Atwood*, 2013, 24). As is evident, such productive occurrences are unlikely to be happenstance.

The series is also part of a larger disruption to a singular (distortive) narrative of Atwood as *literary* author, or even more reductively, as *literary novelist*. The graphic genre is contentious as "literature" and, like science fiction, prone to marginalization by what Ursula Le Guin calls discriminatory actions of "the literary bigots" (Le Guin, "*Year*," 2009). Moreover, Atwood's audience is inauspicious: the text is aimed at ten-year-olds, which itself suggests an understanding of devalued (or less visible) interdependencies and urgent cross-generational need for cultivating ecological awareness. In an era where climate activism is led by teenagers like Greta Thunberg, Atwood's engagement with youth should not be surprising. In fact, Atwood contributed to Thunberg's recent compilation, *The Climate Book* (2022), in the future-oriented final section titled "What We Must Do Now." Atwood's emphasis on the importance of younger generations—as both vulnerable and powerful—is echoed by environmental organizations like A Rocha, who proclaim: "We are preserving *sensitive* habitats and *threatened* species, growing food sustainably and feeding people living on low income, *inspiring school children and training young people*" (https://arocha.ca/, emphasis mine). If, as John Berger observes, "[s]eeing comes before words [and] [t]he child looks and recognizes before it can speak" (Berger, *Ways of Seeing*, 1990, 7), then it makes sense to teach children to see the world holistically and interdependently from the first—and to consider their own place in it.

Indeed, Atwood's focus on the act of learning to *see* by paying attention to *what we look at* is embodied in her spare imagistic poetry, and it is significant that her early career was established and dominated by her work as a poet. Sherrill Grace observes that Atwood's emphasis on "the nature of the self and human perception . . . are central to her work" from her earliest poetry collections. Moreover, she notes that "Atwood conceives of the self not as an individual ego, defining itself against its surroundings, but as a place or entity co-extensive with its environment" (Grace, *Violent Duality*, 1980, 2). There is a kind of holistic vision at play here, informing all of Atwood's writing, not solely her poetry. Cultivating the ability to see the particular in its relationality to the whole of creation is what Walter Brueggemann terms "the scandal of the particular" and can be seen as an essential aspect of humanness (*Message of the Psalms* qtd in Rohr, *Just This*, 2018, 11). It is a poetic vision of particulars: vivid and even stark in detail, sharply defined in order to expand our vision and imagination *beyond* what is seen *through* what is seen. In fact, contemplative looking and seeing has been identified by thinkers like Richard Rohr as the means of "know[ing] that we live in a fully *sacramental* universe, where everything is a pointer and an epiphany" (Rohr, *Just This*, 2018, 11, emphasis in original).

This poetic instinct and call *to see* undergirds Atwood's work more broadly as represented, in part, through the recurrence of birding. Importantly, however, Atwood's emphasis on birding is not merely metaphorical, a way of exploring relationality; it has a material urgency that calls for action and the need to humbly acknowledge the intricate ecological balance of our earth. As she notes in an interview with *Audubon*:

> You can't consider birds in isolation from everything else. If you see them as symptoms, the worse things are for birds, the worse they're going to be for us eventually. . . . If you follow birds, it allows you to get a pretty global view of what's happening to the planet.
>
> (Leber, "Margaret Atwood," 2019)

That we are now officially declared to be in the geological epoch of the Anthropocene is itself a kind of confession—a formal acknowledgement of human ecological impact on an unprecedented scale. Atwood's literary engagement with birding distils this overwhelming confession of complicity into digestible forms across genres, over time, and into the real world.

In terms of conservation, Atwood's work is direct but also invitational. Her literary texts—across genres—call us to engagement with our world writ large and to retain hope for collective action leading to positive ecological change. Atwood aligns her views with John Berger's observation that "[i]t is seeing which establishes our place in the surrounding world. . . . The way we see things is affected by what we know or what we believe"; ultimately, he contends, "[w]e only see what we look at. To look is an act of choice" (Berger, *Ways of Seeing*, 1990, 7–8). Unlike Edward and Sarah in "The Resplendent Quetzal," Atwood's own involvement with birding extends far beyond personal recreation to advocacy, action, and invitation. Instead of being a way to evade authentic relationships as it is for Sarah and Edward, for Atwood birding is a platform where she seeks to inspire others to biocentric accountability and deeper engagement. In addition to her ongoing participation with A Rocha, Atwood engages in the politics of conservation and specifically in birdlife preservation. She leverages her social capital by hosting birding cruise tours as Birdlife International's celebrity birder and is Honorary Co-chair of Birdlife International's Rare Bird Club, comprising long-term major donors who share the organization's goal to "strive for a world rich in biodiversity, where nature and people live together in harmony." As the world leader in bird conservation, Birdlife International has recently launched a youth programme to encourage lifelong ecological commitment to responsible cohabitation of our world (www.birdlife.org/). Clearly, engaging youth in conservation is a priority for Atwood, not only through her *Angel Catbird* graphic novel series but via the real-world conservation organizations she supports. In 2003, Atwood and her late partner Graeme Gibson also founded the Pelee Island Bird Observatory, located at the intersection of migratory routes of songbirds on Lake Erie in Ontario, to encourage traditional birding through research, outreach, migration counts, and practical resources such as a downloadable bird checklist and list of species at risk. It seeks not only to educate but to create personal involvement in conservation efforts—including a focus on intergenerational involvement through youth community activators. Distinctively, Pelee Island Bird Observatory combines birding with literature as evidenced through Springsong, a festival established by Atwood in the early 2000s, as well as writers' workshops and an inaugural writing contest established in 2022. In addition, Atwood remains Honorary Chair of PIBO's board, clearly embodying her values in an interdependent way of living (https://pibo.ca/en/).

Notably, Atwood's status as highly acclaimed international writer provides not only a platform for financial support and global reach of this future-oriented conservation message but also a method attuned to reaching younger generations. Atwood has been a formidable presence on the social media platform Twitter, tweeting regularly about birds, green spaces, and global and local conservation issues to her two million followers, and using her literary celebrity as cultural capital for ecological ends rather than personal gain (@MargaretAtwood). Ever aware of uses and abuses of power, Atwood seems keen to introduce issues and perspectives into social discourse across media and generations, creating collective accountability. As one of Atwood's earliest critics observes, "she is quick to point out that the world we perceive is, in some measure, the world we create" (Grace, *Violent Duality*, 1980, 2). Clearly, Atwood advocates for humanity to capture a vision for creating a better world—to work globally towards restoration and relationality, with increasing understanding of our interdependency. Through her bird conservation advocacy and consistent emphasis on cultivating ecological awareness through literature in multiple genres, Atwood calls us to choose *to look* and consequently *to see* our world biocentrically—from what we might call a bird's eye view instead of solely from an anthropocentric one. It is an increasingly urgent "bird call" well worth hearing and heeding.

References

Atwood, Margaret. *Angel Catbird*, vol. 1. Illustrated by Johnnie Christmas and Tamra Bonvillain. Milwaukie, OR: Dark Horse Books, 2016.

———. *Angel Catbird*, vols. 2–3. Illustrated by Johnnie Christmas and Tamra Bonvillain. Milwaukie, OR: Dark Horse Books, 2017.

———. *Bluebeard's Egg*. Toronto: McClelland and Stewart, 1983.

———. *The Complete Angel Catbird*, vols. 1–3. Illustrated by Johnnie Christmas and Tamra Bonvillain. Milwaukie, OR: Dark Horse Books, 2018.

———. *In Other Worlds: Science Fiction and the Human Imagination*. Toronto: McClelland and Stewart, 2011.

———. *MaddAddam*. Toronto: McClelland and Stewart, 2013.

———. *@MargaretAtwood*, Twitter, Joined July 2009.

———. *Oryx and Crake*. Toronto: McClelland and Stewart, 2003.

———. "The Resplendant Quetzal." In *Dancing Girls and Other Stories*, 161–178. Toronto: McClelland and Stewart, 1977.

———. *Survival: A Thematic Guide to Canadian Literature*. Toronto: Anansi Press, 1972.

———. *The Year of the Flood*. Toronto: McClelland and Stewart, 2009.

Benshirim, Zeke. "Daring to Care: Deep Ecology and Effective Popular Environmentalism." *Harvard University: Sustainability News*, 21 June 2016. https://web.archive.org/web/20230424155930/https://green.harvard.edu/news/daring-care-deep-ecology-and-effective-popular-environmentalism.

Berger, John. *Ways of Seeing*. 1972. London: British Broadcasting Corporation and Penguin, 1990.

Birdlife International. www.birdlife.org/

Bowen, Deborah C. "Ecological Endings and Eschatology: Margaret Atwood's Post-Apocalyptic Fiction." *Christianity and Literature* 66, no. 4 (2017): 691–705. https://doi.org/10.1177/0148333117715252.

Brown, Brené. *The Gifts of Imperfection: Let Go of Who You Think You're Supposed to Be and Embrace Who You Are*. 2010. 10th anniversary ed. Center City, MN: Hazelden, 2020.

Carey, Jonathan. "What's Part Cat, Part Owl, and Out to Save the World? Meet Angel Catbird." *Audubon*, 9 September 2016. www.audubon.org/news/whats-part-cat-part-owl-and-out-save-world-meet-angel-catbird.

Csorba, Esther Susan. "Unlikely Superhero: Margaret Atwood's *Angel Catbird*." *Women Write About Comics*, 18 October 2019. https://womenwriteaboutcomics.com/2019/10/unlikely-superhero-margaret-atwoods-angel-catbird/.

Grace, Sherrill. *Violent Duality: A Study of Margaret Atwood*. Montreal: Véhicule Press, 1980.

Hengen, Shannon. *Margaret Atwood's Power: Mirrors, Reflections, and Images in Select Fiction and Poetry*. Toronto: Second Story Press, 1993.

Leber, Jessica. "Margaret Atwood Insists Birds Matter to Everyone—Whether They Realize It or Not." *Audubon*, 25 November 2019. www.audubon.org/news/margaret-atwood-insists-birds-matter-everyone-whether-they-realize-it-or-not.

Le Guin, Ursula K. "*The Year of the Flood* by Margaret Atwood." Book Review, *The Guardian*, 29 August 2009. www.theguardian.com/books/2009/aug/29/margaret-atwood-year-of-flood.

Méon, Jean-Matthieu. 2017. "Will Eisner and the Evolution of the Graphic Novel." *The Conversation*, 3 March 2017. https://theconversation.com/will-eisner-and-the-evolution-of-the-graphic-novel-73892.

Nischik, Reingard M. "Margaret Atwood's Short Stories and Shorter Fictions." In *The Cambridge Companion to Margaret Atwood*, edited by Coral Ann Howells, 145–160. Cambridge: Cambridge University Press, 2006.

Nussbaum, Martha C. *Love's Knowledge: Essays on Philosophy and Literature*. Oxford: Oxford University Press, 1990. Pelee Island Bird Observatory, 2023. https://pibo.ca/en/

Ridout, Alice. "Temporality and Margaret Atwood." *University of Toronto Quarterly* 69, no. 4 (2000): 849–870. https://doi.org/10.3138/utq.69.4.849.

A Rocha Canada. "Match Your Gift with Margaret Atwood." 2020. https://arocha.ca/portfolio-items/atwoodmatch/

Rohr, Richard. *Just This: Prompts and Practices for Contemplation*. Albuquerque, New Mexico: Center for Action and Contemplation, 2018.

Smith, James K.A. *You Are What You Love: The Spiritual Power of Habit*. Ada, MI: Brazos Press, Baker Publishing, 2016.

Thunberg, Greta. *The Climate Book*. London: Penguin Random House, 2022.

York, Lorraine. *Margaret Atwood and the Labour of Literary Celebrity*. Toronto: University of Toronto Press, 2013.

16 "I Just Can't Get Enough of This Place"

The Gifts and Complications of John Terpstra's Love of Hamilton

Matthew Zantingh

I would like to know where I am.

—John Terpstra, *Daylighting Chedoke*, 2018, 35

This is not mere romanticism for Terpstra—it is a statement about the interrelatedness of material things, the relationship of people and trees and rocks and water and highways and cars and houses and shopping malls and stairs up the [Niagara] escarpment.

—Deborah C. Bowen, "John Terpstra and the Sacramental in Urban Geography," 2002, 198

What Is This Place?

If John Terpstra is right when he claims that "We have no idea where we are" (*Daylighting Chedoke*, 2002, 5), then his informal trilogy of books about Hamilton, Ontario—2002's *Falling into Place*, 2014's *The House with the Parapet Wall*, and 2018's *Daylighting Chedoke*—marks a sustained attempt to correct just such a lack of knowledge.[1] At the very beginning of his 2002 work of creative nonfiction, Terpstra declares, "I am attached to a piece of geography" (*Falling Into Place*, 2002, 15), referencing the Iroquois Sandbar, a collection of accumulated rocks and sand from the last ice age that has shaped a particular strip of what is now downtown Hamilton. What began with an unlikely attachment results in a densely layered and interwoven set of stories, natural history, poetry, beautiful illustrations, and reflections on the modern human relationship to land itself over some 300 pages. However, his sense of attachment and loyalty to this particular urban place inspired him to continue asking questions of this locale. His next two books move upward from the ground itself to the buildings Hamiltonians made out of that clay and stone, then to an exploration of Chedoke Creek, an urban creek that flows from above the Mountain down beside the Sandbar and out to Hamilton Harbour.[2] In a way, Terpstra's gaze moves from the soil up to the heights, assiduously taking in the land that he lives on.

While there may be nothing particularly new about writers falling in love with the places in which they live, Terpstra's Hamilton trilogy presents a unique opportunity to witness one author's changing relationship to a specific locale. What starts as a humble declaration of loyalty to a largely unloved part of a post-industrial city

DOI: 10.4324/9781003366744-16

leads Terpstra in several directions at once: one, towards the past rooted in a place, and the particular ways that it literally and figuratively shapes the present; two, towards the past and present human inhabitants of that place to listen to their stories; three, towards the animal, plant, and even the geological inhabitants of a place to witness the stories that can be heard in a sandbar or a buried urban creek; and, four, towards an active engagement in ongoing dialogue with the place. Terpstra explains this expanding relationship in *Falling Into Place*: "this whole venture into one particular place, one piece of geography, has everything to do with relation. A series of interconnected, overlapping relationships that reach further and further out. Relation and reciprocity" (Terpstra, *Falling Into Place*, 2002, 108). In what follows, I am going to read this set of books as a cohesive collection of stories that are embedded in a very local, particular place, with *Daylighting Chedoke* offering a crucial key to interpreting the earlier two books. Vanessa Watts's theorization of Indigenous "place-thought" (Watts, "Indigenous place-thought," 2013) acts as a keystone for the books, its weight holding the arch of a doorway together in this rich, multi-layered narrative about the relationships between settler, immigrant, and Indigenous humans alongside animal and plant neighbours in one small place.

There are consistent threads in each book that give warrant to my sense the books are a trilogy. Of course, any two books that are focused on the particular features of the same area of a mid-sized city will seem connected, but in Terpstra's case, certain techniques and ideas are present in each book, demonstrating continuity. *Falling Into Place* includes a reference to the fact that Chedoke Creek is leaking sewage into Hamilton Bay (95), a topic that plays a key part in *Daylighting Chedoke*. In *Falling*, Terpstra also undertakes a creek walk with a friend (145–147) that foreshadows a series of walks he will take alongside Chedoke Creek in *Daylighting* (17–23). Near the end of *Falling*, Terpstra draws attention to the bricks of the house he inhabits and the fact that they are made of clay from the area (313), an interest which blossoms fully in *The House with the Parapet Wall* with its discussion of locally produced bricks (32–35). *House* returns to the Iroquois Sandbar, where Terpstra now lives, a move he chronicles in the book (43), but it also takes up the creeks that tumble down the Niagara Escarpment, including Chedoke Creek (82–83), anticipating his focus in *Daylighting*. Lastly, the imagined daylighting of Chedoke Creek that Terpstra includes at the end of *Daylighting* (124–126) echoes the imagined freeing of Niagara Falls from human structures in *Falling* (221–24). To be sure, these shared threads are visible retroactively, yet they all stem from Terpstra's desire to know the land he lives on and in.

Perhaps the best way to see the continuities between these three books is to compare the declarations that Terpstra regularly makes of his attachment to this particular place. *Falling Into Place* begins with a simple statement, but Terpstra keeps trying out different versions of it: "I am attached to a piece of geography—and to the surrounding lay of land" (19), "I am attached to the dance of the landscape at home" (26), "I am attached to the pried-open fist of a sleeper" (31),[3] and "I am attached to what has been undone, unravelled, to the redrawn, remembering lines. I am attached to this pursuit of them" (66). These unabashed yet nuanced personal declarations are amplified by other voices announcing their love for this place. For example, Mrs. Brunton, a woman who grew up on the banks of the Iroquois Sandbar, tells Terpstra "I loved every minute that we lived by the marsh. . . . And I regretted ever having to leave" (152). Terpstra also finds historical precedent for his affection in the form of Lady Elizabeth Simcoe's diary entry of June 11, 1796: "I was so pleased with this place that the Gov. stay'd & dined at Beasley's" (29). These statements function as a motif of love for this particular place, but this motif extends into the next two books. In *House*, Terpstra writes: "I have written about the bar, at length, and from the point of view of personal attachment to a landform that is part of the civic geography, that has taken a lot of abuse, and yet survives. I wrote *for* the bar" (95). And in *Daylighting*, Terpstra modifies his statement one more time: "I just can't get enough of this place" (126). While declarations of attachment to a place do not necessarily lead to attention, care, or deep love for it, in Terpstra's work we see how they might do so. This is, in part, because the land he is attached to is neither an untouched wilderness nor a sublime but inhospitable spectacle. Instead, as he writes in *Falling*, this land is "embracing and open. Debased and persisting" (37). Terpstra keeps returning to his proclamation of attachment because it grows out of an abundant love for this specific combination of people, animals, plants, and land.

Terpstra also employs what I want to call a looping technique whereby ideas or phrases introduced in one section return later in his writing so that there is a kind of echoing or resonating that occurs. Each book contains personal reflections on the city and its past, but these reflections are broken into pieces and interspersed with natural history, poetry, maps or illustrations, and accounts of others' lives or historical events. In *House*, Terpstra describes a species of everbearing raspberries (86), and then will come back to that adjectival phrase some 11 pages later to apply it to his mother "Everbearing. This is another word that springs to mind in relation to Mom" (97). A little later, Terpstra and his wife are visiting the Netherlands and experience their hair blowing in the wind near the sea, an experience that Terpstra links to a painting his mother owns of the island of Texel, blurring personal experience with art (92, 98).[4] Terpstra also includes a story of Esther, a young woman in the nineteenth century who lives in the house of the book's title.[5] This fictional story helps to tie the various pieces of the book together, but it also starts to resonate with the life of Anna, Terpstra's mother.[6] The looping technique helps reduce any potential for a reader's narrative confusion, but it also pushes the prose towards poetry, with each book utilizing imagery, metaphor, and the rich depths of language itself. While *Daylighting* and *Falling* fit more easily together because of their explicit exploration of specific features of the land, the shared narrative strategy and the location of *House* make it part of this trilogy despite its primary focus on human mortality.

Moreover, there is one shared motif in all three books that neatly frames Terpstra's approach to land: the idea of place as a gift. The framing of one's home as a gift is significant because it refuses the Western notion of private property and exclusive ownership, and it opens up a relationship between the giver, the land, and the recipient, the human living in place. This conceptual framing can be linked back to the Christian creation story as Eden is given as a gift to Adam and Eve, and to God's promises to the enslaved Israelites in Egypt of their promised land in Canaan.[7] Halfway through *Falling*, Terpstra reflects on living by the Niagara Escarpment: "I live within these grand proportions, only slightly larger now than I was then. It seems a gift, rather than an insult or challenge, to be physically humbled in this way, a kindness" (186). In reference to his new house on the Iroquois Sandbar, Terpstra writes "if it feels like a gift, is it a gift?" before deciding that his new home is a gift from the Sandbar itself (*House*, 95). After discussing the ambiguous origins of the word Chedoke in *Daylighting*, Terpstra declares "Chedoke is a word-gift" because although it may be a mishearing of the English words Seven Oaks by Indigenous peoples, it is also unique to this place (7). In each of these instances, this locale is seen as a gift to Terpstra, and the nature of that giving means that Terpstra must honour the gift. Therefore, he refuses sole possession of this place, but instead he repays the original act of giving by writing a series of books that present the texture and contours of this place for others.

Just as Terpstra honours the gift of this place and its people, I want to pause to honour the gifts I have received. In doing so, I am following Deborah C. Bowen's lead and declaring my own investments in this particular set of stories by one writer.[8] Terpstra's *Falling Into Place* was a serendipitous discovery while I was completing a Master's degree at McMaster University, located just west of the Iroquois Sandbar. I was delighted to find that someone had written at length about the gentle hill that I was living on at the time. I also share Terpstra's ethnic and religious background, albeit I am one generation further along the immigrant's path as my grandparents left the Netherlands for Canada in the 1950s. I could not shake the sense that Terpstra's book was an invitation into listening to the stories of the places I had lived, and so it became a central thread in my PhD dissertation on stories of urban nature in Hamilton, Windsor, and Toronto, Ontario. I completed this work under the supervision of Daniel Coleman, who shows up as a fellow walker in *Daylighting Chedoke*, and who has also written extensively about Hamilton, Ontario. Near the end of my studies, I was introduced to Bowen, and she has become a valuable intellectual mentor and wise guide for my own career as a literary scholar. In this chapter, I am attempting to add a small contribution to her insightful readings of Terpstra's work across a series of articles. This chapter is not solely my story but instead is intertwined with the stories of several others (many of whom are also included in this volume) and is rooted in the same place we have all come to know and love. My own experience resonates with Terpstra's claim that if you "live anywhere long enough[,] the landscape will claim your loyalty, your heart" (*Daylighting*, 92).

Locating This Particular Place

The particular nuances of Terpstra's transformation as a writer are also timely, given the current grappling with North America's long histories of colonial oppression of Indigenous peoples and ecological devastation. For example, Michi Saagiig Nishnaabeg scholar Leanne Betasamosake Simpson states:

> Over the past two hundred years, without our permission and without our consent, we have been systematically removed and dispossessed from most

> of our territory. We have fought back as our homeland has been stolen, clear-cut, subdivided, and sold to settlers from Europe and later cottagers from Toronto. The last eels and salmon navigated our waters about a hundred years ago. We no longer have old-growth white pine forests in our territory.
> (Simpson, *As We Have Always Done*, 2013, 4)

She explicitly connects the theft of land with environmental harm. She is not alone in seeing tangible connections between colonial and ecological violence.[9]

While Terpstra may be somewhat belated in noting these devastating ties, his attention to the land prompts questions which led him there over time. Already in *Falling into Place*, Terpstra is aware not only of the way land can be broken by human activity but also of the necessity of inhabiting a place in a more sustainable manner. After the 1997 Plastimet factory fire darkened the sky with columns of smoke while also releasing high levels of carcinogenic chemicals into the air and the bodies of firefighters,[10] Terpstra writes, "within its escarpment embrace [the Sandbar] holds our airborne contaminants, the exhausted smudge of industry, internal combustion and constant motion, for us to breathe in daily" (277). He recognizes how industrial processes, the human desire for comfort, and the demand for consumer goods has scarred the earth itself. However, this statement comes in a section that is a prayer, thanking "Great God Spirit Almighty . . . for the wounds on this landscape, that I touch, that I share" (277). Terpstra's commitment to this particular body of land is intense, in spite, or perhaps in part because, of the long history of damage done there. Terpstra notes how this mutilation began from the earliest moments of settlers arriving on the Sandbar to clear the forests and fill in the marshes, linking violence done to the land with the dispossession and displacement of Indigenous peoples. However, in this first book, Terpstra skirts around Indigenous presence, trying to free himself from the colonial image of "the ecological Indian" or some other romanticized vision, but struggles to do so.[11] At a late point, he ponders, "Perhaps when I stand on the Iroquois Bar and yearn to sense a Native presence I'm already getting what I ask for. The presence is the place itself" (277). What he discovers over the next 16 years is that there is an Indigenous presence, and that is most clearly articulated in *Daylighting Chedoke*, which also ties colonial and ecological violence together. What these three books offer, then, is a path to follow for readers, writers, and thinkers, a path modelled above all on humility, careful listening, and attention to a given place.

It is Terpstra's struggle to situate his own immigrant history within a land that has a long history of both Indigenous and settler stories that is useful for contemporary readers. His engagement with Indigenous peoples and philosophy is complicated in *Falling*. Throughout, Terpstra chafes against popular settler ideas of Indigenous life even as he recognizes the potential usefulness of an idea of "the ecological Indian." After attending a Winter Solstice Ceremony, a syncretic blend of Indigenous spiritualities with New Age philosophy, he writes:

> That what we were doing was a version of a native ceremony, conducted and attended solely by non-natives, I took to be just another of the irreconcilable, odd facts of life in the big city, life on the cusp of a new millennium.
> (75)

However, he cannot shake his suspicions at what is playing out, and later asks, "Who's reminding whom of what, here? Who owns, and who's borrowing? Was there really any legitimacy to these solstice proceedings?" (77). Terpstra raises the question of appropriation and the troubling ease with which Indigenous spirituality is repackaged as one of many "alternative spiritual possibilities" for settler consumers (77). Several pages later, Terpstra records his own response to genuine Indigenous spirituality in his poem "Ceremony," a poem written specifically for *Falling Into Place*. In the poem, the speaker and his wife journey down to Hamilton's waterfront and are drawn to the sound of drumming. The speaker is surprised to find three Indigenous people, "drumming/the sun down,/singing rhythmic undulations over/the chanted surface of the bay" (82). At the end of the poem, the speaker returns to these "three first-peoples . . . chanting and drumming/the sunset as an event/in which they were participant,/as we looked on" (82). The speaker may or may not be Terpstra, as his poetry is often conversational and autobiographical. However, it is intriguing that he turns to poetry rather than prose to convey this experience. He positions a poem, conspicuously titled "Ceremony," after his account of a rather different ceremony he took part in. The poem's location suggests that this is Terpstra's answer to the awkward and uneasy politics of Oriah Mountain Dreamer's ceremony. Significantly, the poem situates the speaker and his wife as onlookers rather than as participants in a ceremony greeting the sunset. However, throughout the book, Terpstra undertakes his own versions of ceremony on the land.[12] Terpstra calls himself a "good citizen guest" (22) because he refuses to call himself native to Hamilton (271), instead foregrounding his identity as an immigrant and internal migrant with roots in several places in Canada. Rather than let racial identity politics define him, Terpstra ends the book by defining himself as "both landed gentry and honoured guest" (313), a gesture which acknowledges history in a place but also refuses sole proprietorship of that place.[13]

If *Falling Into Place* is marked by a complex declaration of belonging in Hamilton, *Daylighting* lands in a more comfortable position as Terpstra engages more explicitly with Indigenous thinking. His intent is signalled in an epigraph from the Stó:lō writer Lee Maracle that reads, in part, "I hope [Canadians] fall in love with the land the way I have: fully, responsibly, and committed for life" (*Daylighting,* x). Terpstra is quoting Maracle, *My Conversations with Canadians*, 2017, "Conversation 8: Jack Scott and the left." Maracle's words make space for non-Indigenous people to share in the experience of this land's gift. Several pages later, Terpstra introduces the Mohawk scholar Vanessa Watts's concept of Indigenous Place-Thought:

> The idea is that society is built in conversation with the world that surrounds it. In the interaction between people and place. We make pacts and historical agreements with the earth, the animal world, the sky world, the spirit world, the world of rocks and trees and flowing streams. It is an active thing. A two-way street. Social relations, personal relations. The connection is a sacred one.
> (*Daylighting,* 5–6)[14]

Throughout *Daylighting*, Terpstra will loop back to this idea of a pact with one's place (72). Late in the book he writes, "The Stewards of Cootes Watershed live up to the terms of a two-way pact when their volunteers haul out garbage from the gorges, ravines, valleys and waters of the creeks that empty into Cootes Paradise" (99). This is a significant moment because the descendants of the settlers who began the process of transforming the land into industrial space, actions that poisoned the waters, buried creeks, and radically altered the ecosystem, are now working to reverse that change. Their physical labour of lifting garbage out, piece by piece, honours the pact with the land. Terpstra describes the pact: "The terms of this pact are not written down, but its fine print is in the leaves and rocks, and in the water" (99). Read charitably, Terpstra has been trying to honour this pact between human, animal, plant, and land since he first fell in love with the Iroquois Sandbar. It may have taken two decades to find the concept to give intellectual grounding to his attachment, but the connection is fitting, as Watts works at McMaster and is a member of the Haudenosaunee who live west of Hamilton on the banks of the Grand River. Terpstra had explored the history of these Indigenous people in *Falling Into Place* (273–75), but it took him some time to find the local answers he was looking for.

Although Terpstra now has the language he needs, the path towards ecological restoration and reconciliation with Indigenous peoples is less clear. A few pages later in *Daylighting*, he writes "you wonder how . . . we as a culture can act on this desire [for reconciliation], since we are not just part of the problem but at its root, even if we are seven generations into feeling and enjoying ownership of the place" (101). Terpstra offers no easy answers. Late on in *Falling*, he struggles to articulate his own sense of identity in relationship to place in the form of a prayer: "I thank you that I am not from here, but am here, an immigrant adoptee, a guest who stayed, a loyalist, a stone who came to land, to rest, on this bar of belonging" (277). The flickering between different metaphors and terms arises, I suggest, from Terpstra's discomfort with the language available to him. However, Terpstra's insistent raising of the question of how we belong to this piece of land is essential to his desire to practice loyalty to the place. His disquiet with how contemporary Canadian culture treats the rocks and dirt of Hamilton has led him down numerous paths, but it feels like his discovery of Indigenous philosophy and story in *Daylighting* makes space for a more generous way of being in the land.[15] By naming himself a "good citizen-guest," he highlights the complexity of belonging to a place with a rich history and offers a non-exclusive vision of place where one has many neighbours.

In doing so, Terpstra's approach is marked by both a high level of humility and an unwillingness to ignore injustice. This approach resonates with the prophetic injunction of Micah 6:8: "and what does the Lord require of you but to do justice and to love kindness and to walk humbly with your God" (NRSV).[16] Throughout the Hamilton trilogy, Terpstra's approach is the antithesis of an arrogant, assertive claim to land. His approach lines up more with Cree theologian Raymond Aldred's diagnosis of contemporary Canada:

> There is a need for settler descendants to properly learn how relationality, *wâhkôhtowin*, also determined and *determines* their creation story. This will help us live together in a good land, in right relations between Indigenous peoples and Newcomers. Without a shared story, we can't journey together.
>
> (Aldred, *Our Home and Treaty Land*, 2022, 14)

Terpstra's Hamilton narrative is his own journey to understand his relationship with this particular land and its inhabitants, but it is not just for him. The very act of publishing the work offers this narrative as a gift to readers spread across the Earth. In "Taking Nature Personally," Terpstra explains why he wrote *Falling*: "Sharing is the reason, finally, why the book was written—or any story is told. We're all in this together, after all" (Terpstra, "Taking Nature Personally," 2016, 44). Therefore, these books might serve as the start of a shared narrative between Indigenous peoples, settlers, and immigrants. It is precisely Terpstra's humility and willingness to keep asking questions and revise his answers that makes the trilogy a starting place rather than the final word.

The Path Through This Place

In each of the Hamilton books, Terpstra turns to the past for some knowledge of what has occurred there. In *Falling Into Place*, this means turning to official history in the shape of the histories of the place (dutifully cited at the back of the book), archival documents, and numerous expeditions to the Hamilton Public Library's archives and the Dundurn Castle museum. Terpstra also includes several excerpts from Lady Simcoe's diary of 1792–1796, written when she visited the area with her husband, Lieutenant-Governor of Upper Canada John Graves Simcoe. Terpstra interprets Lady Simcoe's diary entries as modelling a more attentive and responsible way of being in the land: "If it exudes the privilege of her position as wife of the King's representative, it also expresses the sense of privilege she felt on coming into this landscape" (28). Terpstra is aware that this official visit will start a process of dramatic transformation for this area, but he notices how Lady Simcoe feels like a guest in this place. Likewise, for *The House with the Parapet Wall*, Terpstra mines official histories like *The Dictionary of Hamilton Biography* (Bailey, 1981–1999) and Michael B. Katz's *The People of Hamilton, Canada West* (1975). He even states in the closing "Notes" that he "wanted to contribute something to the Hamilton shelf ever since [he] first read Marjorie Freeman Campbell's *A Mountain and a City*" (152). *Daylighting Chedoke* includes references to official histories, but it also includes a historical map of the city of Hamilton from 1880 in its endpapers.[17] Terpstra works to make the past visible in the present landscape of Hamilton because it can give a lively sense of knowing a place, even if he warns "breathing in the past can be hazardous to your health" (*House* 68).

Where history can sometimes feel abstract, Terpstra's trilogy avoids such a fate by consistently weaving personal stories into each narrative. Each of the books is, of course, Terpstra's own narrative of his being-in-place, but he refuses to allow his voice to be the only one. *House* is very much the story of Terpstra's mother Anna and her painful struggle with Alzheimer's. The book is dedicated to Anna (11), and Terpstra seems intent on making his mother's story more permanent by writing it out. Alongside Anna's story are those of his friend Peter's childhood home at the foot of the Niagara Escarpment (128–29), and a chance connection with the previous inhabitant of the house he has just purchased (77). Similarly, *Falling* contains numerous stories from those who lived in or around Cootes Paradise, a bay connected to the westernmost point of Lake Ontario. Terpstra seeks out two Mercer boys who grew up in a house that was later demolished to make way for a multi-lane highway. He takes them back to the landscape of their childhood to hear their stories of living in this place Terpstra has come to love (117–42). This pattern is repeated again in *Daylighting* as Terpstra includes numerous accounts from residents of upper Hamilton, who tell of their knowledge of the now-buried creek (4, 15–17, 47, 94–97). As Doreen Massey argues in *For Space*, "we recognize space as the product of interrelations; as constituted through interactions, from the immensity of the global to the intimately tiny" (Massey, *For Space*, 2005, 9). Terpstra's archiving of local, personal stories falls on the tiny end of the spectrum of relationships that make up part of Hamilton, yet the very act of repeating these stories alongside those of "official" history gives readers a much richer sense of Hamilton as a place. At risk of seeming obvious, a place is its people, and Terpstra charitably extends his narrative space to those who have a story to tell.

What's more, Terpstra also wants to include, as much as he can, the voices of the created world in his stories of Hamilton. Indeed, *Falling Into Place* is the story of the Iroquois Sandbar, a geological feature that is mostly buried beneath a dense urban fabric of roadways, buildings, and parking lots, emerging in small glimpses until it branches out over Hamilton Harbour. The maps included in the book visually locate the sandbar for those not from Hamilton, but Terpstra uses natural history to tell the story of how this particular sandbar came into being and then turns to human history to explain how it has been radically altered almost from the first moments European settlers arrive. He describes the Sandbar as if it can speak to him: "As the landscape continues to explain itself, as it slowly reveals more of its story, it remains a source of wonder to me how much of that story is written by water" (186). In *Daylighting*, he asks why he pursues "this dirty little creek, this insignificant trickle? . . . if the land is *alive and thinking*, then listening to and engaging these living streams could help us retune our relationship with this place. By the song of its strings, we may be healed" (30). Here, Terpstra grants moving water the ability to not just speak but also to instruct and teach humans. However, this "retuning" will require significant work as modern Westerners ignore those voices by default. In this sense, Terpstra's work feels anti-modern, suggesting a re-enchanted world where the past is still present, where animals and plants have voice, and where the land itself shapes us. In *House*, Terpstra bluntly states: "Walls talk. They speak in simple facts of brick and wood and glass" (16). Again, this speech feels jarring, so Terpstra unpacks the history of local brickmaking (34–35), architectural features like a parapet wall (18), and placements of stairwells in order to give one particular house a voice (146–47). Terpstra even asks, "Can brick and mortar love the ones who live within their walls?" (35). While the literal answer is no, Terpstra's poetic writing raises the possibility that perhaps bricks do speak and do feel something. After all, he argues that locally produced bricks are "indigenous to the earth beneath our feet, which are an articulation of this earth," even if he admits this is a "fancy" (35). What is particularly important is that the natural world is speaking to Terpstra in an urban setting, not in some remote provincial park or uninhabited forest. Some critics may query the arrogance in imagining a human can speak for rocks or water, yet I suspect Terpstra's response would be who will speak for these beings then?[18] In Terpstra's eyes, these particular places have suffered silently for too long and what humans desperately need to do is to stop talking and start listening. In a discussion of prayer in *Daylighting*, he writes:

> Simply pray God the Great Spirit, trees, the sky, the brick, the person walking their dog around the park, the trash can. Call them forward into your attention. Attend, with no agenda or plan. No ask. It's not your business to decide what others may need or want.
>
> (80–81)

Here is an expansive community of voices who might speak, if only humans listen.

Nonetheless, listening is not an end in and of itself for Terpstra as the listening leads him towards relationship and ceremony. In all three of the books, Terpstra reports on impromptu or planned ceremonies on and with the land. In *Falling Into Place*, it is walks with his children to try and find a lost pathway (288–89) and a planned picnic meal near Dundurn Castle that recreates Lady Simcoe and her husband's dinner with the Beasley family (306–311). In *House*, it is a bike trip through downtown with a friend that meanders towards the city's football stadium on Labour Day (98–102) and a reflection on the act of pruning his apricot trees as he considers the way that the loss of his mother is like the destruction of his city (143–144). In this latter ceremony, Terpstra's life becomes a tree that is pruned through suffering, and that tree is rooted in the soil and water of the Iroquois Sandbar as he and his mother have been rooted in this place for several decades. In *Daylighting*, the ceremonies include Terpstra's pilgrimage up Chedoke Creek to the approximate site of its spring (38–39) and a lengthy description of a hike taken with friends to the base of Chedoke Falls (115–122). But the book also contains an imagined ceremony whereby Glooskap, a divine figure from the Indigenous peoples of the Wabanaki Confederacy, daylights Chedoke Creek and turns it into a playground for all beings (124–126). This last ceremony can only be imagined because Terpstra senses that it would cost too much to bring the creek back to the surface and not enough people would care to do so (31). He writes: "this meandering line of words may be the only form of daylighting Chedoke sees" (31). Yet this act carries weight in a tangible sense as the very book, not to mention Terpstra's public readings of it, give voice to the creek in a broader human public. The book may be a small gesture, but it is a voice for the long-abused land all the same. What's more, each of these ceremonies, while seemingly insignificant, puts Terpstra in dialogue with the land, the plants, and the animals that live there. He joins in an ongoing dance of living that goes on without us. Or, to use a Biblical metaphor, Terpstra hears what the psalmist describes in Psalm 66:4: "All the earth worships you; they sing praises to you, sing praises to your name" (NRSV).

Throughout these three books, Terpstra frames the land as sacred, even when it is polluted and damaged. This framing is not a move towards some kind of pantheism but an attempt to counteract the damaging way that modernity treats land as either inert or as material to be exploited however humans decide it should be.

In a long section of *Falling* that discusses Terpstra's uneasy encounter with Oriah Mountain Dreamer, Terpstra senses something meaningful in granting land a larger role than just backdrop. He writes:

> The image of Grandmother Earth is strong, though, and powerfully attractive. . . . The earth deserves exactly the same respect and care that the oldest living people to whom you are physically related deserve. Hurting them is quite literally hurting yourself, because you follow them.
>
> (91–92)

Terpstra wants to take up Indigenous concepts of placing ancestors in the land, but he feels uneasy with the potential colonial appropriation that might constitute. At this point, Terpstra reverts to a more orthodox Christian image, borrowing the Welsh poet David Jones's vision of the land as the body of King Arthur instead. Bowen has unpacked this poetic homage in detail,[19] but for my purposes, Terpstra's claim that "there's a connection, here, between that [Christ] figure and the landscape, between the suffering human body and the body of the earth" (94) sets up the land as more than just inert material. It is something to be treated with the greatest care and respect. And that holds true even when the Iroquois Sandbar is an "open wound," and Cootes Paradise is filled with carp and sewage (95). Later in the book, Terpstra makes the bold claim that this particular polluted landscape is his "lectio divina, the spiritual reading of a passage of holy scripture" (280). In making this remarkable claim, Terpstra refuses the lure of unpolluted places or unpeopled wilderness and instead asserts that this broken body/land is a holy text.

Daylighting Chedoke continues Terpstra's attempt to redefine land in a more profound manner, but this time through a strategy of parallel stories of settler and Indigenous peoples. Terpstra explicitly frames his creek hiking with Daniel as "personal, short-chapter versions of El Camino de Santiago de Compostela, the ancient and popular pilgrimage route in southern Europe" (27). In re-applying this ancient Christian practice to the Hamilton landscape, the implication is, again, that this land is sacred: that it deserves the same kind of veneration, care, and attention that the medieval European Christian pilgrimage receives. A little later, Terpstra describes the creek as "a limber, lithe, winding country road to the river's highway" before declaring it "holy" (35). Where earlier Terpstra seems to shy away from this bold proclamation, he shows no reticence here. Part of the reason why he makes this declaration is that he includes both Haudenosaunee and Christian creation stories (34–35, 42–43). In both of these stories, the land itself is created by the hand of God or divine figures, making the land itself an object of divine care and attention. Yet Terpstra has consistently noted how the Christian story too easily leads into neglect of the places we live or to ecological violence (*Falling Into Place*, 72–73). Terpstra also includes a poem about Glooskap in *Daylighting* as a way to try and find a smoother fit between his Christianity and the place he lives (69, 73). However, Terpstra backs away from this usage as he becomes "a little uneasy about the project, and not just because Glooskap is not local to Head-of-the-Lake" (73). In his unwillingness to mix and match Indigenous stories with his own Judeo-Christian ones, Terpstra refuses the kind of postmodern syncretism of Oriah Mountain Dreamer. And yet, Terpstra honours these Indigenous spiritualities by defending them as belief systems that are "real" and not metaphor or invention. He writes, "Let the prayer-walking wilderness guide of my upbringing in the Judeo-Christian tradition, the one who made the map I go by, also be the walking, talking, earthling brother to, if he will, Glooskap. Let them parley" (75). This kind of humble and open approach to faith is a marked refusal to take up the colonial arrogance which for too long has marked western Christianity.[20] Part of the reason why Terpstra includes these stories is because they frame the land as an active partner in living with humanity. Terpstra hopes that the land might lead us towards ecological healing. *Daylighting* ends with a declaration that all he

> really wanted to do is remember [his] first loyalty and keep faith with the creek while pursuing it and being pursued by it, travelling upstream and down through this landscape of valley and highway, culvert and gorge, stone house and suburban development, pollution and vandalism and beauty.
>
> (126)

Again, Terpstra refuses to give up on the land despite the way humans have continuously ignored it, manipulated it with no regard for others, or actively poisoned it. In some sense, this is a Christ-like belief in the land as worthy no matter how broken it becomes.

Finally, the most significant element of Terpstra's trilogy is his lived demonstration that humans must face the place where they live, not turning their attention elsewhere. While Terpstra is a poet and writer, his environmental praxis is to turn his attention to Hamilton, Ontario, and argue that this particular place is worthy of our deep love and care. Near the end of *Falling Into Place*, he writes:

> Perhaps this desire to leave the city and head to the hills or lakes is the desire to escape from a romance suffering from too much history, in which the partners are having difficulty seeing each other as individuals any longer, much less as lovers. In the wilderness there's a new landscape to romance, a new partner, who briefly allows you to forget the one back home, the one who carries the signs of experience on his or her body, the highly unromantic scars of age, ailment, abuse.
>
> (309)

In this elaborate metaphor, Terpstra imagines the landscape as a lover, drawing on ideas he has already explored earlier in the book. However, he pushes the metaphor further and argues that the human desire for pristine nature is a betrayal of the long-term commitment we have made to our home places. Terpstra's metaphor of a Christian marriage, one founded on long-term covenant that binds bride and groom together in front of God and their community, may feel uncomfortable for some readers. However, it is precisely this sense of a long-term commitment which Terpstra wants to get at. Without the sacred underpinnings and divine blessing that inform a Christian sense of marriage, it is all too easy to simply abandon one partner for another. Terpstra writes bluntly: "love is difficult. The practice of love is complex, contradictory, and embattled" (308). He is not a naïve romantic by any means. Instead, he writes, "The habited landscape of this mid-sized city and surrounding countryside understands me, its brokenness mirrors my own. This long-term intimacy with the nature of where I live is both a choice I've made, and a daily decision" (309). Fourteen years after the first book was published, Terpstra would expound his use of this marriage metaphor: "the word *married* denotes a relationship between equals. Personal. One-to-one. A relationship of love and respect, that also happens to make possible, literally, the continuation of life on the planet" ("Taking Nature Personally," 2016, 44). It is easy to see *The House with the Parapet Wall* and *Daylighting Chedoke* as the fruit of this long-term intimacy. In language that echoes his marriage metaphor, Terpstra closes *Daylighting* with a declaration of his "first loyalty and [desire to] keep faith with the creek . . . Now, considering all that Chedoke and the waters of the world must deal with while continuing to sustain life, I'll have to let that faith keep me" (126). The creek is the beloved in this passage, and Terpstra gives it his first loyalty, much as a bride or groom vows to do in a wedding ceremony. Significantly, Terpstra also puts his faith that this relationship will sustain him into the future as it already has done so.

Where to From Here?

For most North Americans, the place where we live is an urban centre. In the 2021 Census, Statistics Canada found that "nearly three in four Canadians (73.7%) lived in one of Canada's large urban centres, up from 73.2% five years earlier" ("Canada's large urban centres"). These centres, each with at least 100,000 citizens, are places just like Hamilton, Ontario. In the United States, the divide is starker, with 80% of citizens living in census urban areas in the mainland United States (2020 Census Urban Areas).[21] For too long, caring for a place was imagined as caring for a place distant from where one lived: in provincial or state parks, in old-growth forests in remote locations, or on lakes in cottage country. Terpstra's work calls attention to the land most North Americans live in. If we take up Terpstra's concept of the land as a gift, then we must treat the gift of our home place with all the love and care we can muster. Not that we all must become poets and pilgrims to a buried creek, as Terpstra has done, but we can certainly turn our eyes to those writers who have done so and follow them towards a richer relationship with the land where we live and its many inhabitants. In doing so, we may live out what Deborah C. Bowen describes in "Listening to the Voice of Creation": "that Christians discover anew the Word made flesh, actively speaking in and through the suffering but still beautiful ecology of this earth that is home to us all" (232).

Notes

1 To my knowledge, Terpstra has not said explicitly that these books form a trilogy. The biographical note in *Daylighting Chedoke* does indicate that the third work is a

"companion book to his two earlier books about Hamilton as a living, breathing geographical location" (135).

2 The Mountain is a reference to the Niagara Escarpment, a hundred-metre-tall cliff that bisects the city into lower and upper parts.

3 This statement draws on Terpstra's love for David Jones's poetry, and his poetic work reviving the medieval English tradition of viewing the landscape as the body of the King. Deborah C. Bowen has discussed this connection in her "Seeing Beyond the Scenery," 2103, 10–11. Terpstra returns to this David Jones connection in *Daylighting*, 111.

4 Another example of this looping pattern occurs earlier where Terpstra recalls his mother saying "Go ahead," allowing him to ride his bicycle as a child despite religious rules against it, and then two pages later, Terpstra writing to his now-dying mother "Go ahead," allowing her to leave this life behind (*House*, 68, 70).

5 Esther's name may be an allusion to Esther, a young Jewish woman living in exile in the Persian empire but chosen by King Xerxes I as his wife for her beauty. I believe Terpstra wants to draw on Esther's sense of living in a place that is not her home but having to make it so, a position shared by her fellow Jewish exiles.

6 In a structural sense, the Esther story helps overcome some of the potential narrative bumps that arise from the fact that the book originally started as life as three-part essay in the online literary journal *Hamilton Arts & Letters*. Terpstra revised those pieces and added a fourth chapter for the book publication.

7 Of course, the gift of Canaan also entailed the expulsion or genocide of the Canaanites who inhabited that land. Walter Brueggemann's *The Land* (1977) explores the idea of land as a divine gift in detail. On a different note, this concept of the land as a gift could also be linked to Indigenous concepts of the land. In *The Clay We Are Made Of*, Mohawk scholar Susan M. Hill discusses the Haudenosaunee's Drum Dance and Thanksgiving Address as ways to "[honour] the gifts of Creation—all of the beings that exist in this world and the gifts they provide to humans" (Hill, *Clay*, 2017, 26). Exploring the connections of Terpstra's work with Haudenosaunee thinking is beyond the scope of this essay.

8 See Bowen's "Listening to the Voice of Creation," 2020, 219–220.

9 See also Potawatomi philosopher Kyle Whyte's "Settler Colonialism, Ecology, and Environmental Injustice" (2018), and Potawatomi scholar Robyn Wall Kimmerer's *Braiding Sweetgrass* (2013).

10 For more on this fire, see Cheryl Lousley's "The Hamilton Plastimet Fire" (1999).

11 See Shephard Krech III's *The Ecological Indian: Myth and History* (1999) for a discussion of this settler myth of Indigenous peoples.

12 For a more detailed discussion of this ceremony, see my "In the City I Long For" (2014), 179–196.

13 Terpstra revisits this metaphor of being landed gentry in *House* as he playfully imagines himself as the duke of his new house, able to enjoy a backyard pool. However, this section ends with the blunt declaration: "there is no entitlement. No one owns anything" (113), undercutting the metaphor by suggesting the house is itself a gift that he will pass on to others.

14 Watts introduces Place-Thought in "Indigenous Place-Thought and Agency," 2013. Kyle Whyte explores similar territory in Anishinaabe thought, drawing on awbanawben, "the interdependence of humans and nonhumans in ecosystems" ("Settler Colonialism," 2018, 127).

15 Although Terpstra does not mention the Two-Row Wampum or the Dish-with-One-Spoon treaty, there are interesting parallels between his refusal to "own" space and his willingness to share it with others. See Daniel Coleman's chapter in this book for more on those treaties.

16 It is worth noting that earlier in this chapter of Micah, the LORD invites the Israelites to make their case before the mountains and the hills, making clear that the created world itself will judge whether the Israelites have faithfully lived out their covenant with God. I suspect Terpstra would approve of this rhetorical move.

17 Two chapters of *Daylighting Chedoke* appeared in an earlier form in *Hamilton Arts & Letters*, and in those digital essays, Terpstra included numerous historical maps, paintings by Edward Roper, and other pieces of visual history. *Falling Into Place* and *House with Parapet Walls* both include illustrations but not these kinds of visual markers of the past.
18 See Min Hyoung Song's *Climate Lyricism*, 60–64, for a defence of anthropomorphism or personification as a productive technique that "invites thought about the divide between the human and the nonhuman" (62).
19 See Bowen's "'Seeing Beyond the Scenery': Exploring the World Through Metaphor," 2013, 11–12.
20 In allowing Wabanaki and Haudenosaunee creation stories narrative space alongside a Christian one, Terpstra may be following the model of the Two-Row Wampum treaty between the Haudenosaunee and the Dutch explorers and settlers of the early seventeenth century. At the heart of that treaty is a promise that each party will remain in its own boat, and together they will float down the river. For a detailed overview of this treaty, see Richard W. Hill Sr. and Daniel Coleman's "The Two Row Wampum-Covenant Chain Tradition as a Guide for Indigenous-University Research Partnerships" (2019).
21 The US Census Bureau defines an urban area as having at least 2,000 housing units or a population above 5,000. This threshold is quite a bit lower than Statistics Canada's definition.

References

Aldred, Raymond and Matthew Anderson. *Our Home and Treaty Land: Walking Our Creation Story*. Kelowna: Woodlake Books, 2022.

Bailey, T. Melville. *The Dictionary of Hamilton Biography*, 4 vols. Hamilton: Dictionary of Hamilton Biography, 1981–1999.

Bowen, Deborah. "John Terpstra and the Sacramental in Urban Geography." *Literature & Theology* 16, no. 2 (2002): 188–200. www.jstor.org/stable/23927023.

———. "Listening to the Voice of Creation: How Contemporary Ontarian Creative Writers Hear the Natural World." *Christianity & Literature* 69, no. 2 (2020): 219–236. https://muse.jhu.edu/article/758807.

———. "'Seeing Beyond the Scenery': Exploring the World Through Metaphor." *Paper Presented at the Centre for Philosophy, Religion and Social Ethics at the Institute for Christian Studies, "Toronto Inter-Faculty Colloquium,"* March 27, 2013. Toronto, Ontario. Published in *Canadian Theological Review* 2, no. 1 (2013): 59–78. Link to talk: https://ir.icscanada.edu/bitstream/handle/10756/346292/Bowen_Deborah_20130327_1.pdf. Link to essay: https://cetactr.files.wordpress.com/2014/05/ctr-2013-v2-1-bowen.pdf.

Brueggemann, Walter. *The Land: Place as Gift, Promise, and Challenge in Biblical Faith.* Minneapolis: Fortress Press, 1977.

Hill, Richard W., Sr. and Daniel Coleman. "The Two Row Wampum-Covenant Chain Tradition as a Guide for Indigenous-University Research Partnerships." *Cultural Studies: Critical Methodologies* 19, no. 5 (2019): 339–359. https://doi.org/10.1177/1532708618809138.

Hill, Susan M. *The Clay We Are Made of: Haudenosaunee Land Tenure on the Grand River*. Winnipeg: University of Manitoba Press, 2017.

Katz, Michael B. *The People of Hamilton, Canada West: Family and Class in a Mid-Nineteenth-Century City*. Cambridge, MA: Harvard University Press, 1975.

Kimmerer, Robyn Wall. *Braiding Sweetgrass: Indigenous Wisdom, Scientific Knowledge, and the Teachings of Plants*. Minneapolis: Milkweed Editions, 2013.

Krech, Shephard, III. *The Ecological Indian: Myth and History*. New York: W.W. Norton and Company, 1999.

Lousley, Cheryl. "The Hamilton Plastimet Fire: A Political Anatomy of an Environmental Disaster." *j_spot: Journal of Social and Political Thought* 1, no. 1 (1999): n.p. www.yorku.ca/jspot/1/lousley.htm.

Maracle, Lee. *My Conversations with Canadians*. Toronto: Book Thug, 2017.

Massey, Doreen. *For Space*. London: Sage Publications, 2005.

Simpson, Leanne Betasamosake. *As We Have Always Done: Indigenous Freedom through Radical Resistance*. Minneapolis: University of Minnesota Press, 2013.

Song, Min Hyoung. *Climate Lyricism*. Durham: Duke University Press, 2022.

Statistics Canada. "Canada's Large Urban Centres Continue to Grow and Spread." *The Daily*, Statistics Canada, 9 February 2022. www150.statcan.gc.ca/n1/daily-quotidien/220209/dq220209b-eng.htm

Terpstra, John. *Daylighting Chedoke: Exploring Hamilton's Hidden Creek*. Hamilton: James Street North Books, 2018.

———. *Falling into Place*. Kentville: Gaspereau Press, 2002.

———. *The House with the Parapet Wall*. Kentville: Gaspereau Press, 2013.

———. "Taking Nature Personally." *The Goose* 14, no. 2 (2016): 43–44. https://scholars.wlu.ca/thegoose/vol14/iss2/41.

———. "Why Must We Die? An Introduction." *Hamilton Arts & Letters* 4, no. 1 (2011): n.p. https://samizdatpress.typepad.com/spring_2011_hamilton_arts/john-terpstra-why-must-we-die-1.html.

———. "Why Must We Die? Part 2." *Hamilton Arts & Letters* 4, no. 2 (2011): n.p. https://samizdatpress.typepad.com/fall_2011_hamilton_arts_l/john-terpstra-part-ii-why-must-we-die-1.html.

———. "Why Must We Die? Part 3." *Hamilton Arts & Letters* 5, no. 1 (2012): n.p. https://samizdatpress.typepad.com/spring_2012_hamilton_arts/john-terpstra-part-3-why-must-we-die-1.html.

United States Census Bureau. "2020 Census Urban Areas: Facts." *United States Census Bureau*, 9 February 2023, www.census.gov/programs-surveys/geography/guidance/geo-areas/urban-rural/2020-ua-facts.html

Watts, Vanessa. "Indigenous Place-thought and Agency Amongst Humans and Non-Humans (First Woman and Sky Woman go on a European World Tour!)." *Decolonization: Indigeneity, Education & Society* 2, no. 1 (2013): 20–34. https://jps.library.utoronto.ca/index.php/des/article/view/19145/16234.

Whyte, Kyle. "Settler Colonialism, Ecology, and Environmental Injustice." *Environment and Society* 9, no. 1 (2018): 125–144. https://doi.org/10.3167/ares.2018.090109.

Zantingh, Matthew. "'In the City I Long For': Discovering and Enfolding Urban Nature in Ontario Literature." PhD diss., McMaster University, 2014. http://hdl.handle.net/11375/15979.

17 Can We Hear What the Land Is Saying? The Haudenosaunee Two Row Wampum and the *Via Negativa* as Postures for Listening

Daniel Coleman

Shortly after I moved to the city of Hamilton, Ontario, in 1997, to a position as a professor in the Department of English and Cultural Studies at McMaster University, I met Deborah C. Bowen, herself a new professor at Redeemer University, a few kilometres across the city. She herself was relatively new, having moved to Hamilton just the year before. We were new professors, both invested in the challenge of learning how to study and teach literatures published in Canada and wondering how our Christian heritage would shape and be shaped by our immersion in the worlds of literary teaching and study. Little did we know, back then, how the place we had moved to would impose itself, become guide and context for what it is that we profess. Little did we anticipate how the beautiful and bruised gift that Hamilton is, with the Niagara Escarpment curving around the end of Lake Ontario and embracing the hyper-industrialized city with its steel factories pouring black smoke into the sky and coal tar into the bay—little did we realize how this place would shape us as students and teachers. Little did we know how central this question—Can we hear what the land is saying?—would become in our ongoing intellectual and spiritual formation.

With some of Canada's richest farmland drained and sheltered by the 600-foot cliff face of the Niagara Escarpment that surrounds Hamilton Harbour at the Head of the Lake, this unique, UNESCO-recognized biome at the western end of Lake Ontario is both one of the most naturally blessed places in the entire country and one of the most abused and disrespected. Like many cities in the industrial heartland of the Great Lakes region, Hamilton is a classic "rust-bucket" city, the result of 150 years of industrial dumping in the watershed, particulate-ridden smoke boiled into the air, and hectares of "brown-fields," where now-bankrupt and razed factories and warehouses have left soil laced with toxins and industrial effluent for future generations to deal with. It is the kind of place that makes you suck in your breath at the beauty that persists, despite our blatant human disrespect. The undeniable and visible contradiction of this place shouts for even the most jaded person's notice, and I suspect that the difficulty, perhaps impossibility, of responding adequately to the clear cries of this desecrated, attractive place makes most of us doubt our eyes and ears, causes us to distrust our ability to respond.

One of the many unlooked-for gifts of this beautiful, damaged place was that my job at McMaster put me in the company of Haudenosaunee (Six Nations)

DOI: 10.4324/9781003366744-17

colleagues and students. The Six Nations are an estimated thousand-year-old Confederation of the Mohawk, Oneida, Onondaga, Cayuga, and Seneca nations, who were joined by the Tuscarora in the 1720s. Their most concentrated location today and Canada's most populated reserve, the Six Nations of the Grand River, is a half-hour drive south of the city. Elders from the reserve had worked with the university to establish an Indigenous Studies Program, which had been in operation for seven years when I arrived on campus. I soon found myself engaged with students and professors from ISP.

Conversations with them turned out to be life-changing, because Haudenosaunee colleagues regularly challenged my English professor's assumption that books—reading and writing—were the central focus of teaching and learning. What else does a teacher do but get students to read books? What else does a learner do but take notes, review pages, and write essays? But even those Haudenosaunee colleagues who loved and studied books like I did, regularly asserted that, ultimately, *the land is our teacher*. Books may help us listen and learn, they insisted, but the land itself is the ultimate teacher.[1]

I had no idea what this meant. I was educated and formed in a settler-colonial culture that teaches us to distance ourselves from land. Following the philosophies of John Locke, my culture thinks of land as something that we must "improve" by transforming it with our intelligence and labour, and once we've improved it, it takes on value.[2] As an object for human use. To us, unimproved land is not just neutral, it's a bit frightening, even alien. It is dark woods or dangerous cliff faces where waterfalls plunge deep into the unknown depth of earth below. Untamed land and its climates can kill you. Settlers tend not to trust the land itself to teach us things, except to be wary of it, and to distrust those who have a centuries-long set of, to us, animistic relationships with that land—who believe land is alive, has spirit and sentience. Through our collective experiences of distrusting both the land and the First Peoples of that land, we learned to distrust ourselves—our own trustworthiness in relation to land and Indigenous people. As a result, to ask what the land is saying, what land is teaching, is not a question our culture has been comfortable in asking: we are pretty clear on what we humans have to say *about* land, but settler colonial culture has not been very interested in—has even avoided the question of—what the land itself has to say.

I should say, right now, that I am a product of settler colonial culture as much as anybody else is, so I am only starting to grope my way along these counter-cultural lines of inquiry and reflection. I hope I can offer some of my early observations, but I will not be able to answer this question in any definitive way. Perhaps, you could think of my reflections as being about how to get in a good *posture* for listening to what the land has to say.

One reason why it's a difficult question for us to address is that most of us were taught to avoid it. "Man versus Nature" was a major theme in the school curriculum across North America.[3] The curricular pioneer story of how settler colonies came to be—in Canada and the United States, and in others like Australia and New Zealand—is built on the blood, sweat, and tears our ancestors sacrificed in the fight to clear the land and make it into what we call "civilization." Refugees from religious wars, from kings and queens and their armies, from plagues and pestilence, from famine and poverty—settlers—have come, generation after generation right up the present, to the Americas to make a new life and society. To do this, they fought against a natural world that was unfamiliar to them: they chopped down the dense forests, they pushed the people who lived here out of the way, they dammed up the rivers to grind flour and then make electricity—not way off somewhere, but right here. The town of Dundas now incorporated into the City of Hamilton was known early on as Dundas Mills because the creeks running over the escarpment were perfect for turning the wheels of many kinds of industry—grist mills, paper mills, lumber mills. So the roots of the steel industry in Hamilton grew out of the industrialization of the escarpment's waterfalls (Freeman, *Hamilton*, 2001, 20–21).

Consumptive Culture

If you were to boil it all down, settler colonial society perceived the land not as something sacred and alive, sentient, autonomous, and expressive, but as inanimate, an object that could be consumed, turned into products and commodities. It's not hard to sympathize with how this mindset developed: people who had been hungry and oppressed needed the land to feed them; desperation made them inventive about how to turn the land into bread, into tables and chairs and tablecloths and sheets, how to transfer earth into consumables. They invented ways to use the energies of earth—firewood, water, coal, steam, eventually electricity—to make more than they needed for our own homes and to export them to places far away. Merchants in the town of Dundas cut the Desjardin Canal through the glacial sandbar that divided Cootes Paradise from Burlington Bay so that products made in the Dundas Mills could seek markets through the Great Lakes, the St. Lawrence River, and eventually the world.

Judeo-Christian ideas were bent into service of this desperate inventiveness: a passage was found, Genesis 1:26, at the start of the Bible where God was reported to have said: "Let us make man in our image, after our likeness: and let them have dominion over the fish of the sea, and over the fowl of the air, and over the cattle, and over all the earth, and over every creeping thing that creepeth upon the earth" (King James Version).[4] This statement was taken to mean that people *should* dominate the earth and all its beings. God had taken soil and made humans from it, so why shouldn't people, made in the image of God, make any number of things out of the land? According to this line of thinking, human inventiveness means improving the things of earth, increasing their value by making them into things we humans need—food, clothing, shelter, and a growing list of conveniences and comforts. Our society became a consumer culture, in which, we today gauge the health of our society by reading the Consumer Confidence Index. If people's capacity to consume is high, then our economy is strong, and we can expect to keep our way of life sailing along blissfully.

But something in us—call it conscience, call it ethics, call it spirit—knows that consumption can also kill you. From 1600 to 1900, one out of every four Europeans died of a disease they called "consumption" (we call it tuberculosis today; see "History of World TB Day"). Nineteenth-century English speakers called it consumption because it was a mysterious disease that ate you from the inside, a lot like we think of cancer nowadays. Your life was going steadily along, and then one day you coughed blood into a handkerchief, and you knew something was consuming you inside. This awareness of consumption as inner disease points up the irony in which we live: in order to live and survive, we have to eat, we have to consume, we have to turn the land into bread, but consumption is also how we die. We ourselves live within its cycle: we will inevitably be consumed by the earth and turned into soil ourselves. We consume, then, to ward off inevitable consumption. Some of our distrust of land, then, some of our distrust of ourselves and our neighbours, has to do with our desire to escape our own place within the cycle of consumption.

Ears to Hear

Living in distrust, with a desire to escape, is not a good posture in which to address a question like the one we are considering here: "Can we hear what the land is saying?" In order to keep my mind steady with this question, I did what bookish people often do: I decided to take notes and write about the literal urban land where I live. Writing is my way to train my attention. Others may take photographs, make drawings and maps, take samples and look at them under a microscope. For me, I look, listen, and take notes, then I look at the sentences I am making to see if they accurately echo what I saw and heard. As soon as I started out, I knew that any attempt to listen to the land would be overwhelming. Even the sheer number of lives and stories living in one square metre of soil—the microbes, root fungi, earthworms, seeds of coming life, decomposition of old life—will quickly blow anybody's mind. So I thought, okay, not something too big or far away, but right here—my own yard will suffice. At least I can watch it all year long. And paying attention to a backyard in this everyday, but amazing city—with its desperate mix of natural beauty and desecrated soil, water, and air—will be familiar to people who don't live in a picture postcard wilderness. The resulting book, *Yardwork* (2017), starts in my backyard, and it attends to the human and geological stories in the soil, the interconnections of our watershed, the more-than-human neighbours in urban pine trees or urban deer. Beyond these, it traces the traffic in hummingbirds, juncos, squirrels, and cicadas that connects my yard—that connects every yard in this city—with Central America and the Arctic Circle, as well as deep underground.

Writing about the place where I live was my effort to take up the counter-cultural question at the heart of this essay, given that I am very aware of how hard it is for twenty-first-century Western readers to re-tune ourselves to the language of earth, the ways in which Creation expresses itself beyond our settler focus on consumption, our way of turning everything around us into commodities and possessions, into food and shelter to sustain our current ways of living, into ideas and arguments that buttress our human centrality, our dominion over everything else.

Urban deer offered one important entry point. Their presence in the city—seven in the family who walked regularly down our street early in the mornings, 14 encountered on one half-hour stroll in the parking lots in the valley below where I live—is a regular, surprising reminder that we share this urban space with large, sentient beings, whose adaptations to post-industrial city living mean their numbers are increasing year to year. I began to try to get to know the doe family who live in the strip of white pine, red oak, black walnut, spruce, and maple woods that creates a wild corridor along the slope between my neighbourhood and the parking lots below. Little seeps of spring water leak out of the slope at regular intervals, giving them sheltered spots where they can drink.

The first thing you realize about deer is that they realize you before you realize them. For one thing, their hearing is better than human hearing. Their tall, white-furred ears are already blazing in your direction, the second your eyes surface over the ridge and you come into their presence. These are urban deer. Unless you have a dog with you, they don't vault away, and may even take a few tentative steps towards you, their round eyes alert, their ears tuning you in. They may stamp a front hoof, and you may think this a nervous tick. Maybe an auditory warning signal, reinforced by vibrations in the ground that other deer can feel. I didn't know until reading up on deer communication that it also works by smell. White-tailed deer have interdigital glands, located between the two halves of each hoof. A sharp jab into the earth releases yellow waxy stuff with a strong odour. What I thought were dainty jabs were olfactory phonemes.

Smell is central in deer communication. Not only do whitetails have scent glands between the toes of their hooves; they also have them on the outside of what looks like their ankle but is technically called the metatarsal joint. They've got another set on the insides of their knees, technically the tarsal or, in horse language, the hock. When I first started watching deer in my neighbourhood, I saw they all have black, leathery-looking patches on the insides of their hocks. If you look at a deer head-on, you'll see they have a slightly knock-kneed stance, especially in the long, slim, back legs. So I had wondered if the black leather patches were calluses where the fur gets worn away by constant rubbing when they walk. Not at all. These tarsal, metatarsal, and interdigital glands are all parts of their communications system. All they have to do is walk and their hoof glands leave wax signs of their passing on the ground. Whenever they drag an ankle through the leaves of a shrub, or when their knees rub each other, they're expressing themselves, leaving messages other deer and animals can read—a whole system of sentience and feeling that exceeds most human people's notice.

So what are the land and its denizens saying? If we want to know what deer are saying, apparently we need not just be people with ears to hear, but also ones with noses for fragrant conversation. We need to know what sense to attend with. And knowing which sense takes generations of intimate attention.

Hungry Listening

The Lakota writer and theologian, Vine Deloria Jr., wrote that lots of settlers can't understand the languages of the birds, animals, and plants that compose the "other peoples" of our ecosystems because we haven't lived here long enough.[5] Indeed, the word *settler* is misleading, because even though our ancestors "settled" in Canada or the United States or Mexico, our people seldom stayed anywhere on this continent long enough to learn the local languages of the people who already lived here, let alone those of deer, or pine trees, or water. Our people moved and moved. Moreover, one of the tragedies of our colonial history is that our ancestors worked very hard through institutions such as the residential school system to kill the knowledges Indigenous people had collected from living here for thousands of years by suppressing their languages and disrupting the intergenerational transfer of knowledge so that many Indigenous people—though luckily, not all—no longer have access to those knowledges either. Our cultural arrogance enabled us to kill a great deal of the knowledge we now so desperately need.

I also think that our ability to hear is distorted by the ever-present desire to turn whatever we consume into a product, a commodity, or an idea we can sell, use, trade, or exchange. Dylan Robinson, a xwélmexw (Stó:lō) professor of Indigenous Arts at Queen's University, says that part of the problem is what he calls "hungry listening." We settlers listen in order to consume. He calls it "hungry listening" after the term for settlers in his Halq'emélem language, which means "hungry people" because the waves of settlers who arrived on the coast of British Columbia, where he's from, were starving when they came off the boats in the desperate, foolhardy goldrush years. The Stó:lō people fed these "starving people" (Robinson, *Hungry Listening*, 2021, 2). Part of the settler colonial problem is that by starving ourselves from a spiritual reciprocity with the land and its peoples, we have stayed hungry ever since.

One of the ways hungry listening can distort what we hear is when we think that by learning to listen like Indigenous people do, we might shed our settler ears and become Indigenous ourselves. This is a very subtle temptation: developments like an increasing awareness of the harm our consumer society has done to the environment, or an awareness of how our Residential Schools consumed Indigenous people in the effort to turn them into settlers, can make us want to slip out of our own society's history and ways of doing things and to ease our way into others'—some of us want to become more Indigenous than Indigenous people themselves, who have lost many of their traditions by means of our society's interventions in the continuity of their ways of life. Some of us want to be the new Indigenous Canadians or Americans—though, of course, without going through the traumatic history of colonization.

We can't hear what the land is saying if we try to be somebody we aren't, by appropriating someone else's culture, by pretending to hear through other people's ears. We need to learn postures and protocols for listening that don't distort what we hear by turning it immediately into fuel for restarting our own engine. For feeding and reinforcing our own sense of entitlement and centrality in the worlds we inhabit.

The Two Row Model of Respect

One of the fortunate things about living here in Hamilton, Ontario, is that one of the most simple and yet profound formulations for developing a respectful way of engaging with others' knowledge and ways of being was invented over four hundred years ago by the Haudenosaunee people, our neighbours who live today here on the Grand River territory. The Covenant Chain-Two Row Wampum is one of the first political-merchant-legal-ceremonial agreements made between incoming Europeans and Indigenous peoples in North America. And right from the start, it addressed the subtle desire to assimilate other lives into one's own. Indigenous leaders such as Onondaga faithkeeper Oren Lyons ("Gä•sweñta' Reflections," 1) and Grand Chief of

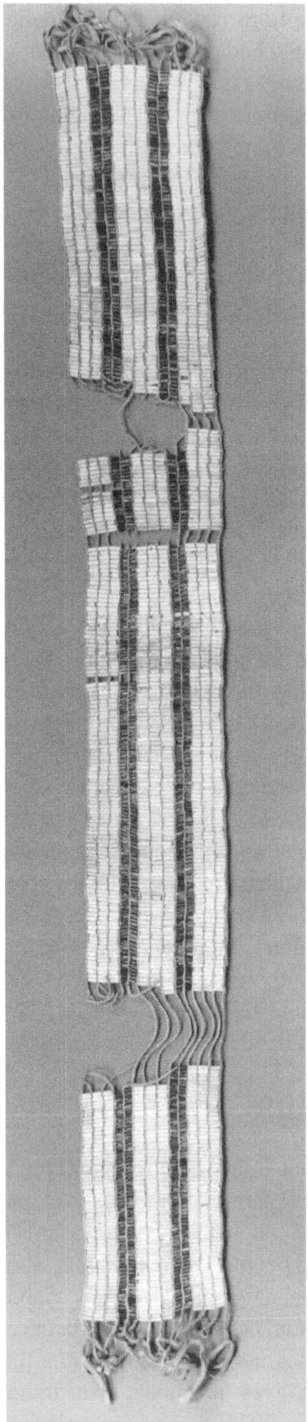

Figure 17.1 Tékeni Teyohà:te, “two roads or paths.” This Two Row Wampum of unknown age is now in the possession of Six Nations Confederacy Council.

Photo courtesy of Rick Hill.

the Assembly of First Nations Ovide Mercredi (*In the Rapids*, 1993, 35) have called it the "grandfather of the treaties" because it created the precedent for treaty-making in Dutch and then British colonial law that, in turn, set the precedents for Canada's recognition of Aboriginal and treaty rights in Sections 25 and 35 of the Constitution Act.

After the Dutch arrived on the Hudson River in 1609, they encountered *Kanien'kehá:ka* (Mohawk people), the easternmost of the Six Nations Confederacy of nations, and over the next years they began to develop a trading relationship that they formalized in about the year 1613. At the negotiations, the Haudenosaunee explained that Creator wanted everyone to live in peace. "The time had come," they said, "for the white people and the Onkwehonwe ["original" people] to continue as friends so that all people may walk upon this earth in peace and love one another."[6] The Haudenosaunee Confederacy had a long history of using shell beads woven into strings and belts in different coloured patterns to symbolize and record important ceremonies and treaties.[7] Observing this Haudenosaunee practice, the Dutch suggested a symbol of their own derived from the chains that anchored their sailing ships: "The symbol of this Covenant is a three-link chain which binds this agreement made by us, and there is nothing that will come between us to break the links of this chain." One name for this agreement, then, has become the "Covenant Chain of Friendship." The Mohawk spokesman took up this three-link chain and gave it this interpretation: "The first link shall stand for friendship, the second will stand for our good minds, and the third link shall mean there will always be peace between us. This is confirmed by us" (Thomas & Miller qtd in Hill, "Oral Memory," 1992, 154). Haudenosaunee confirmation was ceremonially enacted by presenting the Dutch with a wampum that had two purple rows each two beads wide on a backdrop of white beads, each of the white rows itself being three beads wide and echoing the proposed Dutch chain with its three links of friendship, good minds, and peace.

He then explained the symbolism of the Two Row Wampum:

> I have a canoe and you have a vessel with sails and this is what we shall do. I will put in my canoe my belief and laws. In your vessel you shall put your belief and laws. All my people will be in my canoe, your people in your vessel. We shall put these boats in the water and they shall always be parallel, as long as there is Mother Earth, this will be everlasting.
>
> (Thomas & Miller in Hill, "Oral Memory," 1992, 155)

The Mohawk spokesperson did not require that, in order to enter their river country, the Dutch should go to residential school, learn the Mohawk language and how to read wampum, while they abandoned Dutch language and heritage and assimilated into Haudenosaunee life. Instead, the Mohawks recognized there was good in Dutch ways, and that friendship between autonomous equals was the best way to have a relationship with them. The two parties agreed that they would honour this agreement as "long as the sun shines upon this earth, as long as the water still flows, and as long as the grass grows green at a certain time of the year" (Thomas & Miller in Hill, 155). So they established this treaty of friendship, not on a human time scale, not for a certain number of generations, nor for as long as the New Netherlands company shareholders were still running a profit, or even for how long the current Dutch government stayed in power. The land, the river of life on which the vessels floated, was the timeline. The land and its ways of life, like the humans, were understood to be autonomous equals within the agreement. The best way to establish a relationship that could sustain life on the river was to respect the differences evident in Dutch ways of living and understanding things rather than to consume them and board their sailing ship.

When the British arrived in Haudenosaunee territory after taking over Manhattan from the Dutch in 1664, they needed allies among the Confederacy of the Haudenosaunee nations, so they renewed the treaty based on the Two Row. British colonial records for the next 150 years show that they continually invoked this agreement to keep the Six Nations as allies (not subjects) in their conflict with France over access to the fur trade in North America.[8] These principles of reciprocity and autonomy have informed Haudenosaunee relationships with colonial governments ever since, so that, to this day, Six Nations people insist they are not Canadians or Americans, but sovereign Haudenosaunee.

Hungry settler colonial people's inability to live side by side without consuming the other, however, has meant that in the four hundred years since the Two Row Wampum agreement was made, colonists have commandeered the Haudenosaunee canoe over and over again, to the point where their political autonomy and territorial sovereignty was overwhelmed in what became the United States and Canada, and the confluence of the Mohawk and Hudson Rivers where the agreement was signed, let alone the Grand River or St. Regis Rivers where they were moved, now run with heavy metals, blue-green algae, effluent of many kinds, and *E. coli*. The philosophy of respect and reciprocity, of the Covenant Chain of Friendship between the Two Rows, was overrun and disregarded. We can no longer be confident that the grass will turn green, the river will still flow, and the sun come up in the morning.

The Two Row Wampum reminds us that there is (always) more than one agency, more than one set of ears, in any environment, in any ecosystem. Its philosophy arises from Indigenous knowledge—knowledge arising from people's thousands of years of intergenerational observation and experience, as a confederacy of distinct nations, of figuring out how to live peacefully with others within a shared ecosystem. It reminds us in graphic form that if we wish to live on a healthy river, if we wish to hear what the land is saying, we need to get back in our own boat. Meaning, we need to know ourselves, not by consuming the world but by opening ourselves to an awareness of being known by others. This involves the development and maintenance of a good mind, a posture of respect, a deportment of courteous attention to the otherness of others, of the world.

I've heard more than one Haudenosaunee person explain that the Two Row Wampum isn't exactly a treaty. To return to Oren Lyons's phrasing, it's more like a "grandfather of the treaties." It's a ceremony, a way of representing and then renewing a living relationship. To freeze this relationship into a "treaty," signed and dated at a particular meeting on a particular date, is to isolate one transaction from the living, ongoing relationship that continues to this day. As Six Nations Confederacy lawyer Kayanesenh Paul Williams puts it, when the parties meet to renew the relationship, the

> principles are not invented or accepted each time: they are the rules. The individual transaction, viewed out of the context of the entire system that defined and controlled it, loses meaning, effect, and functionality. It is like seeking to understand an animal without considering its habitat or its family, herd or pack: you are left with an isolated specimen. . . . Environmentally, Haudenosaunee understanding of our place in the universe requires us to greet, thank, and acknowledge every other living being as a relative. The meta-narrative of creation teaches us our responsibilities.
>
> (Williams, "The Chain," 9)

Williams speaks here of all of the treaty partners' responsibilities—our responsibilities to know ourselves within the places where we find ourselves, and, perhaps even more, to allow ourselves to be known by the beings that make up those places.

The Two Row Wampum, then, poses a different model for a way of relating to the world around us from the model of consumption on which settler colonial culture is based. It invites us to enter into a reflexive, responsive, respectful relationship with a sentient world, the literal, living world, the "river of life" on which our lives depend. It presents in graphic form an active awareness of sharing the river of life with other beings, Onkwehonwe and more-than-human, who have intelligent eyes, ears, senses of smell, touch, and taste of their own that know the world independent of our perceptions, ways of knowing, and languages. To enter into respectful, good-minded, and peaceful relationships with the worlds in which we live, we from hungry, imperial cultures need to inhabit our own vessels rather than trying to commandeer the canoes of others.

Respect as a *Via Negativa*

For those of us with settler ancestors, then, how do we learn to engage the world in which we live without consumptive, hungry listening? Without appropriating other people's cultural knowledge, without boarding the Indigenous canoe?

In books such as *Living in the World as if It Were Home* (1999) and *The Larger Conversation* (2017), the Saskatchewan-born poet and nature contemplative, Tim Lilburn, probes the question of what the land is saying deeply and profoundly. He too begins with an encounter with deer: "Deer come out of the poplars just as day becomes night; they move in the blue air," Lilburn writes, "One of them jerks up with a look. They see me standing beside the woodpile. They stare. I stare" (*Living* 3). "I am seen straight through (of that, no doubt) but cannot say how I am seen" (*Living* 4).

What is it about deer? Is it their large eyes that seem so curious and alert, but also unthreatening, that allow us to open ourselves to their gaze? In comparison, a wolf's eyes might make us feel afraid. An insect's compound eye looks to us more like machinery than something with a life and consciousness inside. Whatever it is, in my experience and in Lilburn's, we both register this eye-to-eye encounter with the otherness of Creation by being looked at by deer. Something about their beauty, their gentleness, their quivering vulnerability, makes us wonder: what does the deer see when it sees me?

Like Lilburn, I think this kind of encounter makes us aware of an intuitive desire to enter into a reciprocal understanding with the beings of the natural world. He goes on to say: "The desire to feel otherness as selfhood, to be the deer seeing yourself, remains; for me it never leaves, the old residue of Paradise," of the integration of human with nature we inheritors of Western religions imagine Eden was—a prelapsarian integration that was destroyed by what we call the "Fall from Grace." Lilburn says it's good to feel this desire, but it leads inevitably to the limits of our language, of the words with which we try to capture things; it confronts us with our desire to consume the deer. Something that can only turn the livingness of the deer into venison, into a species in our taxonomy of kingdom, class, order, and genus. It "thingifies" the deer, and whatever it sees when it sees us, remains forever out of our awareness.

The Spanish school teacher and poet Antonio Machado wrote an amazing little poem that clarifies what's at stake in encounters like this:

> The eye you see is not
> an eye because you see it;
> it is an eye because it sees you.
> (Machado, *Times Alone*,
> 1983, 143)[9]

The one-way viewing that has characterized settler colonial culture, so confident in its own knowledge, its own religion, its own science, its own way of seeing everything made us totally unconscious of *being seen* by the land, by nature, by deer, and by all kinds of otherness, including other peoples. It sanctioned our ignorance of living in parallel with the other vessels on the river of life.

It taught us dis-respect: think about the word *respect*—its root is *spect* as in spectacle, perspective, spectator (the act of seeing or looking), and then the prefix *re-* adds the idea of looking back or returned looking. Respect therefore carries within it the idea of being aware of being looked at. *Dis-respect*, then means a refusal of having your look returned, of being seen, a posture that pretends it is not being seen. Our society's avoidance of being seen by nature, by others, has fed our culture of dis-respect, which has allowed us to treat other lives as if they were consumables, commodities, resources we can burn at our own discretion.

But every once in a while, we find ourselves being seen by a deer, or by other elements of nature: by the polluted water, by the brown air, by the dying trees, by the plant or animal we accidentally killed. By the children we sent to Residential School. To listen to the land, to allow ourselves to be seen by nature, requires a restoration of something like awe for the livingness of things, filled with their own cognition and dynamic living, that sees us.

I spoke earlier of Robinson's xwélmexw (Stó:lō) term for settlers being "hungry people." Another Stó:lō writer and philosopher, Lee Maracle, wrote:

> From a [Stó:lō] perspective, study ought to move us beyond the relentless reproduction of our cultural bias and remove the filters binding our ability to see beyond this bias. . . . The spiritual objective of study is to transform the way we see, to broaden the field of vision, to inspire us to 'turn around,' to drink in the images both of the world and the imagination.
> ("Oratory," 2007, 70)

There are different models for how to transform our way of seeing, our way of listening and paying attention to whatever the land might be saying.

I'll turn briefly to one that comes from the Western neo-Platonist, Judeo-Christian tradition, because I think one of the temptations of our decolonizing times is to reject our own culture's compromised traditions and to seek elsewhere for helpful models for thinking and analysis. While this can be an important way of showing the kind of respect modelled in the Two Row Wampum, of modelling the awareness of being seen by other people's ways of thinking and being in the world, the problem is, as Robinson points out, that looking to other traditions can also turn readily into appropriating them. We can avoid that kind of hungry listening, though, by searching for postures of respectful listening that lie submerged and often disregarded within our own cultural heritages.[10] The tradition I have in mind has been suppressed in the grand narrative of imperial Christianity: it is the apophatic tradition (to know the perfection of the Divine through silence more than words, through what cannot be said), more readily known as the *via negativa*, the "negative way," or "path of negation" that was formulated and practised, especially by the men and women of early monasticism, and perhaps most famously described in the anonymously written *Cloud of Unknowing* in the late fourteenth century.[11]

Tim Lilburn describes this *via negativa* through the *eros* (longing) to see himself as the deer see him, even as he is aware of the impossibility of this desire. Any words he can apply to the deer are his own labels, drawn from his own human mind and vocabulary. For this reason, they can only feed back to him his own worldview, his own framework for understanding things. The monks who formulated the way of negation had the same experience of wishing to understand what was beyond human words and human understanding, of wanting to fuse themselves into ultimate life itself: they encountered the limits of their own minds, their own words, their own comprehension. They were very aware that human perceptions of the source of all life are usually a mirror of our own image: rather than us humans being made in God's image, God kept being made in ours. This is the human condition, beautifully rendered, in I Corinthians 13:12 in the image of how we find ourselves in this life looking at a dim reflection in a mirror, and hoping that in the next life, we will be seeing, not through our own reflection, but face to face. "Now I know in part," St. Paul writes, "but then shall I know even as also I am known" (King James Version).

Isn't this what we long for when we ask, "Can we hear what the land is saying?" To know as fully as we are known? To see ourselves as the deer—and all the other beings of Creation—see us?

The *via negativa* was not popular to a Christianity that became distorted with the imperial ambition to consume the world. To improve it and turn it into value, into property. So the negative way was shoved into the margins of the history of the institution of the Church. This, because it taught a humility in the face of not knowing and a dim awareness of being known, being seen by otherness, by the forces in and around us that make life go on. This humility became the way itself for the early monastics: "Contemplation," Lilburn explains about this way of retraining the human mind, "does not wish to subdue the world but to dwell in it" (*Living* 32). "Its knowing is poor . . . Contemplation is the moment when human knowing, lured by the possibility of perfect understanding, is thwarted, shamed, bent back on itself" (32). "This knowing in the midst of the embarrassment of knowledge topples into adoration" (33).

Adoration is a very different approach to the world than consumption. It is like a very amped-up form of "respect," which carries within it not just the idea of having our gaze returned, but also of the kind of esteem that arises from an awareness of the eye that sees us. Adoration or esteem changes the balance of power between the looker and the deer: it recognizes beauty, spirit, sacredness, sovereignty, capacity in the eye that sees me. It adores that eye and the amazing fact of its life beyond my own: it remembers that it is an Eye because it sees me.[12]

This reversal is key if we are to learn a posture in which to listen to the land within which we live. It's not rocket science or even super-mystical: if we want to learn *from* something (as compared to learning *about* something), we have to respect it.[13] We have to believe that it has its own life and its own autonomous way of being, and only then can we give it the kind of respectful ear that isn't just imposing our own philosophy, our own mental map, onto anything the land might say.

I am grateful for the opportunity to reflect here on the question "Can we hear what the land is saying?," because the answer to the question is contained in the question itself. It is a counter-cultural question. It is a question that settler colonial culture has been keen to avoid. But the growing awareness of the harm that consumptive culture has done to the ecosystems within which we live and upon which we depend is changing our minds, retuning our ears, causing us to wonder what it is that we have ignored, what we have been taught to ignore. So, in asking this question here, we are placing ourselves in a different posture towards the world around us, and that posture itself is crucial if we are to make ways of life that last as "long as the sun shines upon this earth, as long as the water still flows, and as long as the grass grows green at a certain time of the year." I am hoping that I've been able to sketch here a couple of models, readily available to us, right in the places where we live, though suppressed within the cultural traditions we settler colonials have inherited—in the model of the Two Row Wampum, the "grandfather of the treaties," and in the Western, anti-imperial religious philosophy of the *via negativa*—for learning to hear what the land is saying.

Notes

1 "Land," writes Kanien'kehá:ka (Mohawk) scholar Sandra Styres, "refers to the ways we honor and respect her as a sentient and conscious being. Therefore, in acknowledgment of the fundamental being of Land, I always capitalize Land. I have come to know Land both as a fundamental sentient being and as a philosophical construct" ("Literacies," 2019, 27). Increasingly, Western scientists are coming to greater understanding of the sentience of nonhuman beings. See the work of Suzanne Simard on the intelligence of mycorrhizal networks between funghi and tree roots in *Finding the Mother Tree*, 2021, for example, or James Bridle's overview of myriad forms of intelligence in *Ways of Being*, 2022, especially Chapter 1 "Thinking Otherwise" and Chapter 2 "Wood Wide Webs."

2 See John Locke's influential formulations, in *Two Treatises on Government*, 1689, on how orderly society generates civil law by making reasonable agreements among people about the transfer of inanimate property. What makes anything into property, Locke

explained, is human labour. A person's right to an oak tree's acorns corresponds to how many acorns he can gather, just as a person's right to fence off land as his own corresponds to how many acres he can cultivate and reap (the masculine pronouns are part of the assumption here). For Locke, intelligent human labour, the cultivation of acorns or planting of corn, constitutes improvement of the land, and this improvement generates the right to its resources, whether that be water derived from digging a well or diamonds dug from under the earth. It is the investment of labour that makes wine, bread, or silk worth more than water, acorns, or skins. And it is the investment of labour that makes fenced and farmed fields worth more than forests and untilled meadows. (See, for example, Locke, section 41, p. 122.)

3 This curriculum theme may seem like an outdated artefact of the mid-twentieth century now replaced by a more integrated understanding of human relations in ecosystems. But it continues to appear in school curricula, updated as "human versus nature" at CommonLit, an online database of curricular resources "that has everything teachers and schools need for top-notch literacy instruction" (see www.commonlit.org/en/themes/human-vs-nature), or the Standards Aligned System (SAS), developed by the Pennsylvania Department of Education (see the Lesson Plans on "Man vs. Nature" at www.pdesas.org/module/content/resources/19460/view.ashx).

4 See Lynn White's famous criticism of the influence of this passage on Western culture in "The Historical Roots of Our Ecologic Crisis," *Science* 115, no. 3767 (10 March 1967): 1203–1207.

5 "Non-Indians have simply not been on this continent very long," Deloria writes in *God Is Red* (1978); "their families have rarely settled in one place for any period of time so that no profound relationship with the environment has been possible. Additionally, non-Indians have engaged in senseless killing of wildlife and utter destruction of plant life. It is unlikely that they would have understood efforts by other forms of life to communicate with humans" (278).

6 I am taking the words quoted in this paragraph from Cayuga chief Jacob Thomas and Onondaga chief Huron Miller's oration of the agreement as transcribed in Rick Hill's "Oral Memory of the Haudenosaunee," 1992, 5.

7 See for an overview, Tehanetorens (Ray Fadden), *Wampum Belts of the Iroquois*, 1999.

8 The version of this agreement made between the Haudenosaunee and the British referred most often to the way the Six Nations had "linked arms" with the Dutch and then British, having tied the European ships to the shore with rope and wampum, and then to symbolize the durability of the relationship linked chains. This version of the agreement was symbolized in what has become known as the "Covenant Chain of Friendship." Historian David Richter finds over a dozen references to the Two Row-Covenant Chain agreement in British colonial records between 1671 and 1691 ("Rediscovered Links," 1982), Jon Parmenter lists 18 of them between 1656 and 1744 ("Meaning of *Kaswenta*," 2013), while Robert Venables indicates that between 1613 and 1842 more than four hundred negotiations were carried out by the Haudenosaunee with the Dutch, the French, the English, and the United States, and that "163 wampum belts associated with these negotiations have been identified and photographed" ("Polishing the Silver Chain," 2011, 3). Lenape legal historian Robert Williams Jr. concludes: "The Covenant Chain functioned as a multicultural constitution in North America from 1677 until the middle of the eighteenth century. The scope of the Chain's influence within British America during this time was extensive . . . Iroquois influence through the Chain, either direct or indirect, was palpable throughout virtually the entire eastern North American diplomatic system" (Williams, *Linking Arms*, 1997, 122).

9 The pun on "eye" and "I" works better in English translation than it does in Spanish with the word "ojo."

10 Very often we can learn where to look within our own cultural inheritance for what has been suppressed or marginalized by seeing it expressed more clearly in other traditions. This is the value of the dialogical or comparative iconography of the Two Row Wampum: it pictures difference, not in order to assimilate it but to clarify what lies in each vessel's own "belief and laws." I have learned to see the value of the contemplative

tradition of the *via negativa* within my own Christian inheritance, for example, by seeing a different version of it developed and practised in Buddhist tradition, for example.

11 "Thought cannot comprehend God," writes the anonymous author of *The Cloud*. "And so, I prefer to abandon all I can know, choosing rather to love him whom I cannot know. Though we cannot know him we can love him. By love he may be touched and embraced, never by thought" (Anonymous, *The Cloud of Unknowing*, 1973, 54).

12 "Because my presence is historically, and actually in this moment, colonial," Lilburn writes in *The Larger Conversation*, "the land may step back before me—or, to say this more accurately, conditions in me will cause it continuously to recede. Nevertheless, my abjectness serves me here, a protean emptiness offering space for approach, and moreover I increasingly sense in dim outline what would meet my need; further clarity comes as confidence within heterodoxy grows. I hang around, loitering with loose intent, wait and see what might come forward. I maintain, as a far as I am able, a non-Romantic, apophatic anticipation: desire propels me into earth apprenticeship" (Lilburn, *Larger Conversation*, 2018, 204–205).

13 "When we practice humility and gratitude," writes the Jain monk and environmentalist Satish Kumar, "we are able to learn much *from* nature. But we in the anthropocentric, modern civilization learn *about* nature. There is a great deal of difference between learning 'from' nature and learning 'about' nature. When we learn 'about' nature she becomes an object of study, leading to exploitation of her . . . But when we learn 'from' nature we establish a close relationship with her. Then there is implicit humility and reverence towards the mystery of natural processes" ("Three Dimensions," 2017, 140).

References

Anonymous. *The Cloud of Unknowing*. Edited by William Johnson. New York: Doubleday, 1973.

Bridle, James. *Ways of Being: Animals, Plants, Machines: The Search for a Planetary Intelligence*. New York: Farrar, Straus and Giroux, 2022.

Coleman, Daniel. *Yardwork: A Biography of an Urban Place*. Hamilton, ON: Wolsak & Wynn, 2017.

———, Kaitlin Debicki and Bonnie Freeman, eds. *Deyohahá:ge: Sharing the River*. Unpublished manuscript submitted to Wilfrid Laurier University Press [in preparation 2023].

Deloria Jr., Vine. *God Is Red: A Native View of Religion*. 1973. 30th anniversary ed. Golden, CO: Fulcrum Publishing, 2003.

Freeman, Bill. *Hamilton, A People's History*. Toronto: J. Lorimer, 2001.

Hill, Rick. "Oral Memory of the Haudenosaunee: Views of the Two Row Wampum." In *Indian Roots of American Democracy*, edited by José Barreiro, 149–159. Ithaca, NY: Akwe:kon P/Cornell University Press, 1992.

"History of World TB Day," *Centres for Disease Control and Prevention*. Updated 12 December 2016. www.cdc.gov/tb/worldtbday/history.htm.

Kumar, Satish. "Three Dimensions of Ecology: Soil, Soul, and Society." In *Spiritual Ecology: The Cry of the Earth*. 2nd ed., edited by Llewellyn Vaughan-Lee, 134–147. Point Reyes, CA: Golden Sufi Centre, 2017.

Lilburn, Tim. *The Larger Conversation: Contemplation and Place*. Edmonton: University of Alberta Press, 2018.

———. *Living in the World as If It Were Home: Essays*. Toronto: Cormorant Books, 1999.

Locke, John. "Chapter V: Of Property." In *Two Treatises on Government*. From *The Works of John Locke, A New Edition, corrected, in Ten Volumes, Prepared by Rod Hay*, vol. V, 115–126. London: Thomas Tegg; W. Sharpe and Son, 1823.

Lyons, Oren. "Gä•sweñta' Reflections." Excerpts from the writing and speeches of Oren Lyons, Onondaga Nation Faithkeeper, assembled and edited by Rick Hill. In *Deyohahá:ge: Sharing the River*, edited by Coleman, Debicki, and Freeman. Unpublished manuscript submitted to Wilfrid Laurier University Press [in preparation].

Machado, Antonio. *Times Alone: Selected Poems of Antonio Machado*. Translated by Robert Bly. Middletown, CT: Wesleyan University Press, 1983.

Maracle, Lee. "Oratory on Oratory." In *Trans.Can.Lit.: Resituating the Study of Canadian Literature*, edited by Smaro Kamboureli and Roy Miki, 55–70. Waterloo: Wilfrid Laurier University Press, 2007.

Mercredi, Ovide and Mary Ellen Turpel. *In the Rapids: Navigating the Future of First Nations*. Toronto: Viking, 1993.

Parmenter, Jon. "The Meaning of *Kaswentha* and the Two Row Wampum Belt in Haudenosaunee (Iroquois) History: Can Indigenous Oral Tradition be Reconciled with the Documentary Record?" *Journal of Early American History* 3 (2013): 82–109.

Richter, Daniel. "Rediscovered Links in the Covenant Chain: Previously Unpublished Transcripts of New York Indian Treaty Minutes, 1677–1691." *Proceedings of the American Antiquarian Society* (1 January 1982): 45–85.

Robinson, Dylan. *Hungry Listening: Resonant Theory for Indigenous Sound Studies*. Minneapolis: University of Minnesota Press, 2020.

Simard, Suzanne. *Finding the Mother Tree: Discovering the Wisdom of the Forest*. Toronto: Allen Lane Canada/Penguin Random House, 2021.

Styres, Sandra. "Literacies of Land: Decolonizing Narratives, Storying, and Literature." In *Indigenous and Decolonizing Studies in Education: Mapping the Long View*, edited by Linda Tuhiwai Smith, Eve Tuck and K Wayne Yang, 24–37. London: Routledge, 2019.

Tehanetorens (Ray Fadden). *Wampum Belts of the Iroquois*. Summertown, TN: Book Publishing Co., 1999.

Thomas, Jacob and Huron Miller. "Oral History of the Two Row Wampum." Cited in Rick Hill, "Oral Memory of the Haudenosaunee: Views of the Two Row Wampum." In *Indian Roots of American Democracy*, edited by José Barreiro, 149–59; 154. Ithaca, NY: Akwe:kon P/Cornell University Press, 1992.

Venables, Robert. "Polishing the Silver Covenant Chain: A Brief History of Some of the Symbols and Metaphors in Haudenosaunee Treaty Negotiations." November 2008. Prepared at the request of Chief Irving Powless, Jr. Updated 11 July & 4 October 2010; and 26 February & 10 March 2011. http://honorthetworow.org/wp-content/uploads/2013/03/Venables-on-the-Covenant-Chain-of-Treaties.pdf.

White, Lynn, Jr. "The Historical Roots of Our Ecologic Crisis." *Science* 115, no. 3767 (10 March 1967): 1203–1207. https://doi.org/10.1126/science.155.3767.1203.

Williams, Kayanesenh Paul. "The Chain, Naturally Understood." In *Deyohahá:ge: Sharing the River*, edited by Coleman, Debicki, and Freeman. Unpublished manuscript submitted to Wilfrid Laurier University Press [in preparation].

Williams, Robert A. Jr. *Linking Arms Together: American Indian Treaty Visions of Law and Peace, 1600–1800*. New York: Oxford University Press, 1997.

18 To Dwell Ecologically

The Practice of Re-Enchantment

Doug Sikkema

Several years ago, my wife and I along with our four children (and one Labrador Retriever) moved to a small acreage in the small farm community of Binbrook, just outside of the city of Hamilton where we had lived in its postwar suburbs for over a decade. The move in conjunction with my doctoral research on American literature, ecology, and theology forced me to confront several uncomfortable ironies: How might I care for this tangible, physical place through something as abstract as academic work? How could I champion the importance of local places and local stories while writing as a Canadian about American literature? And of course: what does my Christian faith, after all of this, *actually* even mean for how I—how we—will dwell in this place?

Growing up in Southern Ontario in the Niagara region within a largely Christian, largely conservative, largely agrarian community, I experienced the environmental movement as something often associated with leftist politics, Buddhist/pantheist spiritualities, or socialist economics—all of which were highly suspect. How we were to relate to the earth—or, rather, the creation—was talked about in church on Sundays in abstract, theological language about stewardship as mandated in Genesis, or by interpreting nature as the "second book" of God's self-revelation. That the Bible talked about the natural world was clear, but how—or *if*—it impacted our actual practices in Vineland, Ontario, in the twentieth century remained unarticulated and, to me at least, unclear.

In retrospect, I realize that the Christian community of my youth rarely, if ever, addressed the methods of our farming community—one which largely relied on heavy pesticide and insecticide use, industrialized farming technologies, and chemical fertilizers. Waterways could be contaminated and the air made toxic by our practices, and even when increasing numbers of our community were diagnosed with infertility or lymphoma cancers (termed "farmer's cancer" by local doctors), many in the community continued to deride environmentalists for their progressive political bent, so-called socialist economic assumptions, and misguided pagan spiritualities of the earth. We did not see ourselves as culpable for what was happening to the ecosystem we inhabited, nor did we see our religion as a resource that had much in the way of offering practical advice for how we might imagine the world and our place in it differently. What, if anything I wondered, did our religion have to say about our responsibilities to *this* environment, to *this* place?

DOI: 10.4324/9781003366744-18

When I started graduate school at the University of Ottawa, I was introduced to Wendell Berry through his novel *Jayber Crow* and guided through it by a wonderful scholar and teacher, Dr. Dominic Manganiello. Wendell Berry, I learned, is a Kentucky poet and farmer who from the late 1960s until today has created a large body of poetry, essays, and fiction about his small town of Port Royal, Kentucky (fictionalized as Port William). Even though rural Niagara and rural Kentucky are very different regions geographically, socially, and historically, there was something in Berry's writing that resonated with my own experiences of small-town Ontario. I began reading everything Berry wrote and soon found that his rural experience was, in many ways, quite similar to mine. And he wrote explicitly as a Christian uneasy with cultural expressions of the faith that are often naïve and complicit with (even participating in) the destruction of the world.

Retaining his membership in the tradition of Christian faith, Berry is ruthlessly critical of supposedly "Christian" gazes fixed on some abstract paradise "beyond" this world. Such forms of religion pay no attention to—and thus give no proper or responsible affection for—the material, embodied, carnal life of particular places inhabited and stewarded by particular people in particular times. In one of his Sabbath poems, "How to be a Poet," Berry advises poets of the land, and "to remind himself," that "There are no unsacred places;/there are only sacred places/and desecrated places" (Berry, *Given: Poems*, 2006, 18). Berry's words struck a chord, debunking my unarticulated assumption that restoration of the environment was something taking place elsewhere or, more likely, in another future time (likely when God came back to restore the creation to himself). Berry's words, however, were profound in their simple practicality. There are no holier, more sacred places than others. Port Royal or Vineland or Binbrook are intrinsically sacred, with or without human use. This insight is inextricably linked to Berry's understanding that the world—every square inch of it—is God's ongoing creative utterance. Any place is capable of being loved and revered and served well, or of being dishonoured, ignored, abused, and desecrated. It's all how we *respond* to the language of the creation poem being sung—or chanted—all around us.

In many ways, what I learned about yard work and Christianity—for better and for worse—was cultivated in Niagara, by my family and larger community. But to have your own house and property is to enter into new forms of responsibility. This place is now, for whatever time we have it, one that I want to know and care for. I want it to become our home.

There is a small creek which runs through our property. It is actually a drainage swale from about four hundred acres of surrounding fields. I looked on the map to see if this creek had a name or fed any larger bodies of water. It turns out that the creek is nameless but is one of many tributaries that eventually leads to Chippewa Creek, which leads to the Grand River, then to Lake Erie and on to Lake Ontario and, finally, out to the Atlantic Ocean. My son, Jackson, decided one day to name our tiny contribution to the great oceans "Frog Creek," in honour of the hours he spends catching bullfrogs and tree frogs there.

I was struck by this innate human desire we have to name things. What is it that makes us see water trickle with the force of gravity and to provide a name for it?

In our first year here, we bought animal and insect guides and downloaded apps to identify tree species by bark and leaf patterns. We learned that the trees in our backyard were actually English Oak and Shagbark Hickory, Red Maple and Curly Willow and Trembling Aspen. The birds who lived here or who were just migrating through were orioles, Cooper's hawks, nuthatches, finches, and even barn owls. There were coyotes and minks, possums, and racoons. There were names everywhere.

The fossil poetry of names made me recall "Breakage," a well-known poem by Mary Oliver. She writes of broken shells found by the sea shore in the "morning light":

> It's like a schoolhouse
> of little words,
> thousands of words.
> First you figure out what each one means by itself,
> the jingle, the periwinkle, the scallop
>   full of moonlight.
> Then you begin, slowly, to read the whole story.
> (Oliver, "Breakage," 2003)

To name the place and its inhabitants was a way of becoming enmeshed in the words of our home, which is ultimately what ecology—*oikos logos*—is all about. Reading the "whole story" of the mysterious names all around us that will always precede and proceed our short span of usage was a way of becoming embedded within the whole invisible community of people who shared these names. Some of the names were ours, a sign of our unique relationship with "our" place. But most other names were like old growth forests of the mind, names passed down for centuries that helped identify the plants and animals who shared our surroundings. This was their home too.

But to name a place is not to know it. At least not completely. We also learned that some of the species (like Red Maple and Wild Grape Vine) were invasive and destructive. Truly knowing their names was a kind of burden, demanding a different form of response. The gypsy moths with their iridescent, dusty wings were beautiful to behold, but terrible in their destructive potential. A beautiful English Oak was completely defoliated in late August. To name the world was only the beginning of our response-ability. What does it mean when "Frog Creek" bursts its banks (as it did in our first Fall) and the soil erosion threatens your neighbour's home? What happens to the plant life under the domineering canopy of the Red Maple or the tangled mess of Grape Vine that pulls Shagbarks down with a potent indifference? What does it mean when the stagnant waters of the creek in July grow a bubbly film on the surface because of the chemical sprays of the upstream farmers and their RoundUp-ready GMO crops?

That the land is not simply something we speak to, but something we hear, is to acknowledge the world as a particularly enchanted place. The beauty of it—given with such wild abandon—arrests our attention and humbles us in all its miraculous complexity. There is such a profusion of life forms in constant states of change. Each season unfolds layers of life and entangled relationships to which we are simply ignorant, bumbling witnesses. The harmony implicit in ecology is a dim echo of the paradise for which the world was intended. Yet the world is a body, broken. The earth groans under its desecration—at our hands or the hands of others. It cries out. But one needs the ears to hear and the eyes to look and see. The creek near our home is awash in the chemical fertilizers and pesticides that run off from the surrounding industrial farms. The forests ache under the neglect that allows grapevine to strangle and topple mighty Oak trees and soft, pulpy Birch. And learning to listen and then hear some of these cries leads one to a distinctly human choice: do we partake in healing or destructive work?

In the past year, we have repaired a crumbled foundation, installed a new septic system that was polluting the waterway, regraded the entire property, and sowed new grass to stop erosion. We continue to look for ways in which life—not just ours—might thrive here, building owl boxes and bat boxes. As we did much of this work, I was struck particularly by the writing of Marilynne Robinson and her Calvinist rumination that the entire world—even our little speck of earth in Binbrook—is part of the vast theatre of God's glory. As a Christian—even one with Calvinist disposition—I tried to see our little place in the world through this particular lens. What does it mean to live, I wondered, *as if* this were true? More, what did it mean to live here on this property as if it were not just an idea but a fleshed, placed, carnal reality? It was hard. I want to see this land as *mine*.

My grandfather was a superb gardener. He passed that on to my mother and, I think, to most of her children. He died over a decade ago now, and the other day my wife and I drove up to the house where he and my "Oma" lived for decades. The gardens he had were the stuff of fantasy and legend in my young mind. But when we drove up to this home, the entire yard was in disarray. The house was shabby and unkempt. The gardens were wild and overgrown. There was no sign of life or love there at all. It was chilling to see the waste and void to which any place returns when human attention and affection are removed.

To experience a place as part of God's theatre is in a way to reconfigure your own work as a responsive participation in the work God is always, already doing without you. The world is no less God's theatre in my grandfather's neglected yard than it was in its glory years when he was diligently stewarding his home. Such a theocentric view forces us to imagine the world as a place that can do without us. It is to allow our deaths to give shape and form to our life. This home I now call mine was, strangely enough, unknown to me not even three years ago. Whatever damage and good I inherit from its previous tenants need my response here and now. Whatever damage or good I do will be passed on to someone else. But ultimately in a century, who knows what will become of this place? It might, like my grandfather's be overtaken by weeds and neglect. It might be stewarded by some great-grandchild. It might be bulldozed to create a new development or incinerated under the falling of bombs from some war lurking behind the veil of the future. I don't know.

But if the world is God's theatre, the troubling fact of my work's seeming futility also comes as a certain form of relief. It is not to say that it makes one optimistic but rather that it circumscribes the hubris with which we—with which I—am tempted to approach my relationship to my place: as if homemaking on this strange wild planet in all the aeons of time all depended on me and my family. As if the world's value was contingent on what humans do to it. As if we could even know that what in our best intentions was not bringing about some unforeseen harm. Understanding work as the stewarding of God's theatre is simultaneously to dignify the power of work and to humble it in the acknowledgement of its final, ultimate powerlessness. Work is the way we practise our re-enchantment and dwell ecologically as members of a place, but it is always an invitation to join the song of creation. As Rowan Williams has suggested, the creative word of God is not simply a song but a summons. And the summons invites us to come. To come and work and pray.

The world's ongoing desecration today creates a certain visceral uneasiness, a certain malaise, in the Anthropocene era in which we live. To understand the world as God's relieves that malaise through a new myth of re-enchantment. It gives us a different story from which to perceive, experience, and act in the natural world. The world, from this story, is to be approached as a gift. The world exceeds our human grasp and calls us into affectionate work to care for it within the radius of our competence and ability. It demands a particular response. Re-enchantment is not—or should not be—an abstract notion but a set of particular practices borne out of an entire form of life, shaped by a story.

However, such a Christian myth of re-enchantment leaves one with a different malaise. The uneasiness comes with the realization that humans are not, finally, the centre of it all. All of us, even with our homes and our lineages, are just so many pilgrims passing through. The home is nothing but a road. We are constantly in motion towards a world in which we are no longer present. This leads one, ineluctably, to the cross-pressure of belief in the secular age. Because all paths eventually lead us beyond all the familiar landmarks of existence and experience and towards an unknown of what (if anything) lies beyond mortal limits.

As I watch my children work alongside me in our tiny plot of time in the twenty-first century, I am preparing them, ultimately, for a world in which I am not. In fact, I am also preparing them for a world in which they are not. I don't know in 50 years, let alone one hundred, what will become of this place, myself, or even them. It's unnerving, to be sure. As I type these final notes, the leaves have (once again) fallen, they are golden and scattered and dying. My children and I have gathered some of them and, with some water and some topsoil, we're hoping that they will compost and be the start of our vegetable garden next spring. What we get out will be a matter of what we put in. This is a basic truth, Berry says, of agriculture. No culture survives without attending to this simple, mundane miracle.

What we put in gives rise to the life that comes out.

References

Berry, Wendell. "How to be a Poet." In *Given: Poems*. Berkeley: Counterpoint, 2006.

Oliver, Mary. "Breakage." *Poetry* (August 2003): 278. www.poetryfoundation.org/poetrymagazine/poems/41917/breakage.

19 Afterword

Deborah C. Bowen

"What has literature to do with environmentalism?" This is almost as dichotomizing a pairing as the traditional one of intellectual Athens with religiously observant Jerusalem in the minds of some Christians, or the arts with the sciences, or scholarship with activism, in the minds of some traditionalist academics. Literature, they might say, is a contemplative and solitary matter, where environmentalism is by definition active and communal. Literature requires personal and private time, where environmentalism requires a conscious inhabiting of social space. Reading literature is a learned and even arcane skill; environmental awareness is something you will perforce pick up these days from almost any exposure to the mass media. Some might even say that reading literature is at best a hobby, if not an indulgence, whereas an involvement with environmentalism is, at this point in world history, clearly a duty.

And yet. One contemporary environmental scientist, Madhur Anand at the University of Guelph, who is also an acclaimed poet and writer of creative nonfiction, argued a few years ago in a talk at the University of Waterloo that "poetry, like chlorophyll, is a catalyst for turning light into energy" (Anand, "Tomorrow Started Yesterday," 2010). There is even a 2009 book titled *Can Poetry Save the Earth?* in which the author, John Felstiner, argues that indeed it can: "For sure, person by person, our earthly challenge hangs on the sense and spirit that poems can awaken" (Felstiner, *Field Guide*, 2009, 357). And this is no mere romantic idyll harking back to an untouched natural world. In his 1992 Nobel Prize acceptance speech, Derek Walcott says of his West Indian heritage, "I have felt from my boyhood that I had one function and that was somehow to articulate, not my own experience, but what I saw around me" (Walcott, quoted in Felstiner, 335). Why? Because there is a "selfhood," to use Felstiner's word, to Walcott's native islands that both pre- and postdates colonialism, and that needs to be reclaimed apart from politics and economics—seen afresh by and for its inhabitants.

Meanwhile, the premier "nature poet" in Canada, Don McKay, has tackled the arts-sciences divide head-on, arguing eloquently for their beneficial interdependence. He writes:

> Geology or, broadly speaking, natural history of any kind, brings the rigour of the scientific frame; poetry brings the capacity for astonishment and the

DOI: 10.4324/9781003366744-19

> power of possibility—or, perhaps more accurately, legitimizes them. Geology inhibits the tendency, most common in romantic poets, to translate the immediate perception into an emotional condition, which is then admired or fetishized in preference to the original phenomenon—fossil, bird, lichen, or landform. For its part, poetry cultivates the astonishment that naturally occurs in the presence of such marvels.
>
> (McKay, "Ediacaran and Anthropocene," 2011, 16–17)

McKay calls for the kind of "poetic attention" that he defines as "a sort of readiness, a species of longing which is without the desire to possess" (McKay, *Vis à Vis*, 2001, 26). This form of knowing "gives ontological applause" by allowing whatever is seen in nature just to be itself, without human intervention; "poetic attention is based on a recognition and a valuing of the other's wilderness" (*Vis à Vis*, 27). This cultivation of astonishment and recognition of what McKay calls "wilderness," by which he means "the capacity of all things to elude the mind's appropriations" (*Vis à Vis*, 21), may be more vitally important than we yet realize. The editor of the lovely creative arts journal *Image*, James K.A. Smith, proposes that,

> In these fraught times, perhaps our greatest need is a renewed imagination. We believe the poetry and painting in *Image* speak to some of the most fundamental hungers and hopes of our world. Now more than ever, we need better ways to imagine how to be human. We need better ways to see our neighbors.
>
> (Smith, "Hungers," 2022)

And seeing those neighbours will include seeing the other-than-humans in our natural environments, as well as the people around us.

Liane Miedema Brown has described in her essay in this present collection how, during the Covid pandemic, I've been collating an anthology of poetry local to the area around Hamilton, Ontario, which renews our imagination about the environment in which we live by acknowledging and preserving the specificities of the place and its nonhuman inhabitants. This region, the area in southern Ontario between the tip of Lake Ontario and the watershed of the Grand River, is shot through with the complex history of both Indigenous and, much more recently, White settlement, and has traditionally supplied fertile agricultural land as well as easy access to water routes. A region often maligned as uninteresting, it has instead been described by one of the poets in the anthology, Adam Dickinson, as "just fascinating" in being a site of the industrialization of the Great Lakes, the displacement of Indigenous peoples, and the ongoing influx of immigrant communities from a shifting global population, as well as including "an international border with the last remaining superpower on the planet—this place is a kind of Ground Zero."[1] It happens also to be an area that is rich in creative writers. Some 40 poets, including John Terpstra, have contributed to the anthology, writing about everything from land and water to birds and trees to farming and food production.[2] My co-editors and I have been struck by the ways in which poetry can indeed make the reader stop and pay attention, and by the fact that attention is the first necessary step towards action. After all, if literature can enable us to see and feel not only the glories but also the present plight of the natural world, it can wake us up to our human responsibility for creating that plight. This is a major part of what it means, now, to enliven our human imaginations in the Anthropocene.

Like the present volume, our poetry anthology, provisionally entitled *Poetry in Place*, is informed by spiritual concerns. Rowan Williams, the former Archbishop of Canterbury, has said:

> Our present ecological crisis, the biggest single practical threat to our human existence in the middle to long term, has, religious people would say, a great deal to do with our failure to think of the world as existing in relation to the mystery of God, not just as a huge warehouse of stuff to be used for our convenience.
>
> (Williams, *Tokens of Trust*, 2007, 50)

Williams is speaking from a Judeo-Christian perspective, which affirms the value of the natural world as God's good creation, given to all of God's creatures to enjoy. But the poems of *Poetry in Place* are not necessarily seeing through a Christian lens. In fact, all the world's great religious traditions have strong ecological beliefs, such that, whether or not a specific religion would describe the natural world as the direct creation of a god, it is incumbent on human beings to act in it with care and responsibility. Already in the fifth century BC "the Buddha taught that all things are interconnected, that the health of the whole is bound to the health of every sentient being. If you harm rivers, trees, animals, soil, you harm yourself" (Steinauer-Scudder, "Hallowed Ground," 2019). Islam teaches:

> Allah is unity, and His unity is reflected in the unity of humanity and nature. We must therefore maintain the integrity of the Earth, its flora and fauna, its wildlife and environment. Our responsibility is to keep balance and harmony in Allah's creation.
>
> ("What Does Islam Teach About Ecology?", Alliance of Religions and Conservation)

Hinduism, yet more expansively, teaches:

> All living beings are sacred because they are parts of God, and should be treated with respect and compassion. This is because the soul can be reincarnated into any form of life. . . . Even trees, rivers and mountains are believed to have souls, and should be honoured and cared for.
>
> ("What Does Hinduism Teach About Ecology?", Alliance of Religions and Conservation)

And many Indigenous religious traditions the world over subscribe to the idea of a Creator, Great Spirit, or Great Mystery—a power or being that has created the world and everything in it (Smith, "Religion and Spirituality," 2018). In North

America, the Assembly of First Nations declares, "If we listen from the place of connection to the Spirit That Lives in All Things, Mother Earth teaches what we need to know to take care of her and all her children" (*Honouring Earth*).

This awareness of the spiritual within all living beings is, then, not a tenet of Christianity's alone, and it was important for us in the *Poetry in Place* project to use a wide-angle lens, because this enabled the natural world to be seen from a number of different spiritual perspectives. But an understanding of environmental responsibility is particularly powerful for Christians, who may even read the suffering of the earth through the suffering of Christ. I am very struck by the fact that several papers in *Questions of Stewardship* affirm this interpretation—I think particularly of the chapters by George Piggford and Karen Dieleman. Of course, the entire volume, as Katherine M. Quinsey eloquently argues in the introduction, is energized by a specifically Christian understanding of environmental responsibility, something that has been brought much more clearly to the front burner for many Christians in the past decade, not least because of Pope Francis's eloquent advocacy of the ethics of an "integral ecology" in his 2015 papal encyclical *Laudato Sí: On Care for Our Common Home*. At the same time, in this volume are many pieces that illustrate Don McKay's notion of poetic attention, which in fact Tina Trigg, in her essay, sees as involving what she calls "a sacramental understanding of the world." This notion can be applied not just to poetry but to other artforms as well and is just as applicable to the past as to the present century.

Thus, Ben Faber takes us back to seventeenth-century England to explore the ecologically freighted beginnings of "paradise lost," and in this same early modern period, Quinsey turns us to theological literature that already defines nonhuman animals as autonomous beings, "created not for the service of man but to glorify God in their own natures." Where Miedema Brown reminds us that we are all members of an ecosystem and exhorts us to recall that "our earthiness is not a taint but . . . the foundation from which we build the Kingdom of God," Susan Emily VanZanten's exploration of a Ghanaian poem witnesses to the fact that this Kingdom is already "inherent in the natural world and its creatures because of God's creative work" and that there are "permeable boundaries between divine transcendence and the natural world." In a similar way, Norm Klassen celebrates a definition of creatureliness that includes both a delight in the immanent and an opening onto the transcendent, while Tina Trigg shows how Margaret Atwood's practice of "following the birds" involves challenging the human perspective and shifting it literally to a bird's eye view of the world. Samuel Martin's creative nonfictional piece overtly refers to McKay, as he describes two artists who struggle to love a place that has been radically damaged by flooding which has destroyed much of their life's work. John Van Rys, in both his short story and his commentary upon it, wrestles with the responsibilities of "creature care as part of creation care." Daniel Coleman encourages us to look to the "immersive listening" of centuries-old Indigenous relationships with the land and to the Christian apophatic tradition, to "learn to hear what the land, what Creation, and the people who built their cultural ways through dialogue with it, might be saying." Matthew Zantingh echoes these concerns in celebrating John Terpstra's attention to the "Place-Thought" of the abused and broken lands and waters of Hamilton. Where

Erin Goheen Glanville's commentary on refugee narratives suggests "a restorative connection between the suffering human body and the body of the earth," Piggford discusses Flannery O'Connor's vision of the world specifically as the suffering body of Christ, and points to that call of Pope Francis for an ecology which recognizes that "social and environmental justice are integrally related." Dieleman shows how poet Alice Meynell sees "the incarnated, crucified, and yet living Christ as continuing to inhabit or wait within or walk among the grain, grass, grapes and galaxies of the world, devoted to their particularity." And Doug Sikkema concludes that rediscovering the world as a place enchanted by God and filled with the glory of his presence constitutes a call to acting responsibly within it.

Rediscovering the world, as we have found in putting together *Poetry in Place*, means rediscovering that the world speaks—that nature has a voice. Perhaps, then, a renewed imagination could lead us to what Robert Macfarlane has called "a Counter-Desecration Phrasebook that would comprehend the world—a glossary of enchantment for the whole earth, which would allow nature to talk back and would help us to listen" (Macfarlane, *Landmarks*, 2016, 32). In these days of pandemic and environmental crisis, that glossary of enchantment seems key. David Abram wrote eloquently during the early days of the pandemic about the insight that "our real collective flesh is that of the Earth itself, this immense spherical metabolism in which we're all embedded and embodied" (Abram, "Ecology of Perception," 2020). And Bruce Epperly says in his book on the present Pope's namesake, Francis of Assisi, "Recognizing that we are truly one in spirit and flesh with all creation, we are inspired to move from self-interest to global concern" (Epperly, *Walking with Francis*, 2021, 31). The writers in *Christian Environmentalism and Human Responsibility in the 21st Century* are urging us to recognize our connections to and responsibilities within the natural world as deeply spiritual issues, directing us away from merely commodifying the world or valuing it only through what McKay calls "the desire to possess." They are recognizing the power of literature to awaken deeper awareness of the world, in such a way that the reader may even feel called to take action. It is surely my hope that, by enchanting their readers afresh through the visionary glossaries of literature, the pieces in this collection will indeed inspire a renewed awareness of our responsibilities within and for the earthly environment in which we have been placed and to which we are all so profoundly attached by divine love.

Notes

1 Adam Dickinson, in interview with Deborah C. Bowen, August 19, 2021.

2 The working title of this anthology, as yet unconfirmed, is *Poetry in Place: Poetry and Environmental Hope in a Southern Ontario Bioregion.*

References

Abram, David. "The Ecology of Perception." *Emergence Magazine*, 2 August 2020.

Anand, Madhur. "Tomorrow Started Yesterday." *TEDx Waterloo Talk*, 25 February 2010. www.youtube.com/watch?v=YGKwMNlYgmU.

ARC: Alliance of Religions and Conservation. *What Does Hinduism Teach Us About Ecology*? http://www.arcworld.org/faiths97af.html?pageID=5.
———. *What Does Islam Teach about Ecology*? http://www.arcworld.org/faithse030.html?pageID=6.
Assembly of First Nations. *Honouring Earth*. https://web.archive.org/web/20230330135008/https://www.afn.ca/honoring-earth/.
Dickinson, Adam. Interview with Deborah Bowen, 19 August 2021.
Epperly, Bruce. *Walking with Francis of Assisi: From Privilege to Activism*. Cincinnati, OH: Franciscan Media, 2021.
Felstiner, John. *Can Poetry Save the Earth? A Field Guide to Nature Poems*. New Haven: Yale University Press, 2009.
Francis. *Laudato Si: On Care for Our Common Home*. Boston: Pauline Books, 2015.
Macfarlane, Robert. *Landmarks*. Toronto: Penguin, 2016.
McKay, Don. "Ediacaran and Anthropocene: Poetry as a Reader of Deep Time." In *The Shell of the Tortoise: Four Essays & An Assemblage*, 9–24. Kentville, NS: Gaspereau, 2011.
———. *Vis à Vis: Fieldnotes on Poetry & Wilderness*. Kentville, NS: Gaspereau Press, 2001.
Smith, Derek G. "Religion and Spirituality of Indigenous Peoples in Canada." *Canadian Encyclopedia* online. www.thecanadianencyclopedia.ca/en/article/religion-of-aboriginal-people. Accessed 19 April 2018.
Smith, James K.A. "The Hungers and Hopes of the World." *Image: Art, Faith, Mystery*. Online appeal, 9 December 2022.
Steinauer-Scudder, Chelsea. "Hallowed Ground: A Profile of Martin Palmer." *Emergence Magazine*, 28 April 2019. https://emergencemagazine.org/story/hallowed-ground/
Williams, Rowan. *Tokens of Trust: An Introduction to Christian Belief*. London: Canterbury Press, 2007.

Index